AF862750

Ecology and ethology of fishes

Developments in environmental biology of fishes 1

Series Editor
EUGENE K. BALON

Ecology and ethology of fishes

Proceedings of the 2nd biennial symposium on the ethology and behavioral ecology of fishes, held at Normal, Ill., U.S.A., October 19-22, 1979

Edited by
DAVID L.G. NOAKES AND JACK A. WARD

Reprinted from *Environmental biology of fishes* 6 (1), 1981
with addition of selected material from the symposium

Dr W. JUNK PUBLISHERS THE HAGUE - BOSTON - LONDON 1981

Distributors:

for the United States and Canada

Kluwer Boston, Inc.
190 Old Derby Street
Hingham, MA 02043
U.S.A.

for all other countries

Kluwer Academic Publishers Group
Distribution Center
P.O. Box 322
3300 AH Dordrecht
The Netherlands

This volume is listed in the Library of Congress Cataloging in Publication Data

ISBN 978-90-6193-896-5 (series)
ISBN 978-90-6193-821-7 (this volume)

Copyright © 1981. Dr W. Junk Publishers, The Hague

All rights reserved. No part of this publication may be reproduced, stored in a retrieval system, or transmitted in any form or by any means, mechanical, photocopying, recording, or otherwise, without the prior written permission of the publishers, Dr W. Junk Publishers, P.O. Box 13713, 2501 ES The Hague, The Netherlands.

Preface

This collection of articles is a sampling of papers from the second biennial symposium for the 'Ethology and Behavioral Ecology of Fishes'. This conference was also the first North American Meeting of the 'International Association of Fish Ethologists' and was held October 19-22, 1979 on the Illinois State University campus in Normal, Illinois, U.S.A., the site for the first conference in 1977 and the proposed site for the third biennial conference scheduled for May 1981. The organizational committee for these conferences is Jack Ward, Chairman, Illinois State University (Biology, Normal, IL 61761); Jeffrey Baylis, University of Wisconsin (Zoology, Madison, WI 53706); Luther Brown, George Mason University (Biology, Fairfax, VA 22030); Mike Fitzsimmons, Louisiana State University (Zoology, Baton Rouge, LA 70893; Mary Henry, Iowa State University (Animal Ecology, Ames, IA 50010); David Noakes, University of Guelph (Zoology, Guelph, Ontario NlG 2Wl); William Rowland, Indiana University (Biology, Bloomington, IN 47401); Benoni Seghers, University of Western Ontario (Zoology, London, Ontario N6A 5B7); Chris Thoms, Mt. Mercy College (Biology, Cedar Rapids, IA 52402); Richard Wyman, Lawler Matusky & Skelly Engineers (Pearl River, NY 10965).

There were 189 registered participants for the second conference and a total of 49 papers by almost 100 authors were given in a single plenary session over a three day period. Each morning was devoted to a special symposium; afternoons consisted of contributed papers.

The symposium 'Schooling of Fishes', organized by Ben Seghers featured invited papers by: Lawrence M. Dill, Robert L. Dunbrack & Peter E. Major; R. Wassersug, M. Potel & L. Katz; Ann C. Hurley; Brian L. Partridge; Gene S. Helfman; Benoni H. Seghers; Lillian E. Roth & Benoni H. Seghers; Scott L. Ralston; Terrance M. Lim; Eva J. Heczko & Faye D. Algranati. The symposium 'Environmental Toxicants and their Effects on the Ethology and Behavioral Ecology of Fishes', organized by Mary Henry and Jack Ward featured invited papers by: John B. Sprague; Mary G. Henry & Gary J. Atchison; as well as contributed papers by Robert A. Drummond, Daniel Abel & Christopher C. Koenig; C.G. Prewitt & C.B. Stalnaker; Kenneth S. Lubinski; J. Doerzbacher & C.F. Bryan; Michael A. Barry; Allen E. Bingham & Robert G. White.

The symposium 'Reproductive Tactics' was organized by Luther Brown, Jeff Baylis and David Noakes, and featured invited papers by: Jerry F. Downhower, Luther Brown, R. Pederson & G. Staples; Roger Thibault; George Barlow; Richard Borowsky & John Diffley; Jack Ward & J.I. Samarakoon; J.R. Shute, P.W. Shute & D.G. Lindquist; James W. Grant & Patrick W. Colgan; B.D. Steel & W.D. Pearson; Wallace Dominey; Frank Meriwether & William L. Shelton.

General contributed papers were presented by: Colin Barnett, Eric Brinsfield & Melvin T. Huish; Lorraine Madsen & John Michael Fitzsimmons; Kathleen S. Cole; Robert A. Bachman; George C. Mitchell & Carl Quertermus; John M. Green & William David Martin; Gerard J. FitzGerald; Martin R. Meisler, Joseph S. Balsano, Ellen M. Rasch, Paul J. Monaco, Edward J. Randle & Kristine Kucharski; Miles H.A. Keenleyside & Brian F. Bietz; Tim Shears; Lin J. Kozlowski; Lawrence M. Page; G. Thomas Chandler & David Lindquist; John S. Stephens & Kim Zerba; Carol Luckner; Harm P. Gross & Dierck Franck; John Janssen; Drew B. Buchanan, A. Bradley Eisenberg & Gerard R. Joswiak; Jean-Guy J. Godin; Michael H. Figler; Kathy Fusner Vail; Paul M. Bronstein; Don Townsend; Bill R. Heugel; Heather V. Roberts; David L. Bechler; Robert E. Muller; R. Craig Sargent, Joseph B. Gebler, Michael Brown & George C. Williams.

All who presented papers had the option of submitting manuscripts to be considered for this special issue. All papers in this issue were subjected to the normal editorial review process. This collection represents about one-third of the papers submitted and a reasonable cross-section of the contributions to the conference. Our selection ranges from detailed biochemical studies to the most wide-ranging theoretical considerations in the evolution of reproductive strategies, from the most recent innovative technology so basic to the study of fish schools to discussions of community relationships. We have placed the papers in a sequence from schooling behavior, to courtship and mating, parental behavior, and finally to community relationships. These do not correspond to the major symposia organized for the conference, but rather emerged as themes with the final selection of papers.

It is perhaps the best indication of the state of interest and activity in this field that we had so many contributors to the conference, and such an overwhelming number and diversity of manuscripts from which to make our final selection. It was indeed an embarassment of riches, and we think it portends well for the future, both of this continuing biennial conference, and for investigations on the environmental ethology of fishes.

We would like to thank all the authors who contributed to this issue for their cooperation and patience during the review and revision process. The following individuals reviewed one or more manuscripts: Faye Algranati, Thomas Ardelt, Gary Atchison, Colin Barnett, Jeffrey Baylis, Richard Borowsky, Luther Brown, David Chiszar, Patrick Colgan, George Constantz, Victor DeVlaming, Wallace Dominey, Robert Drummond, James Grant, Harm Gross, Eugene Helfman, Mary Henry, Lin Kozlowski, Miles Keenleyside, Boyd Kynard, David Linquist, George Losey, Carol Luchner, Mark Matus, William Moore, Randy Moss, Brian Partridge, Lawrence Page, Robert Phillips, Carl Quertermus, William Rowland, J. Samarakoon, Benoni Seghers, John Sprague, Carrie Stafford, Roger Thibault, Richard Wassersug and Richard Wyman.

We also thank the College of Continuing Education, Illinois State University for making all the physical arrangements for the conference. In particular, we give special thanks to Dean Edward Anderson, and Olive Aldrich and Marie Malinauskas. We also give our thanks to an untiring force of students at Illinois State University who made all aspects of the conference run smoothly. They included Ron Allmand, Tom Ardelt, Tom Bell, Kathy Bohn, Darin Collins, Marian Girardi, Ann Gould, Shamelle Grabill, Douglas Hart, Mark Lamon, Mark Matus, Elaine McGuffin, David Meyers, Vicki Miller, William Prynnes, J. Samarakoon, Sue Schaus, Carrie Stafford, Daryle Waechter and Mark Zorn. They acted as projectionists, receptionists, taxi drivers, hosts and hostesess.

David L.G. Noakes
Jack A. Ward

Env. Biol. Fish. Vol. 6, No. 1, pp. 7-13, 1981

A new stereophotographic technique for analyzing the three-dimensional structure of fish schools

Lawrence M. Dill, Robert L. Dunbrack & Peter F. Major[1]

Department of Biological Sciences, Simon Fraser University, Burnaby, British Columbia V5A 1S6, Canada

Keywords:
Photogrammetry, Nearest-neighbour distance, Bearing, Elevation, Coho salmon

Synopsis

A technique using two downward-directed 35 mm cameras has been modified to measure the three-dimensional structure of fish schools. The resulting stereo pairs of photographs are analyzed, producing the 3-coordinate location of each fish's nose, after correction for lens distortion and refraction. Separation angles (bearing and elevation) and distance can then be determined for any pair of fish in the school. The technique's high level of accuracy is demonstrated for an underwater calibration field. It is then applied to the measurement of the 3-D structure of schools of coho salmon (*Oncorhynchus kisutch*) swimming in a hatchery trough. Although the fish were not organized in a rigid crystal lattice, the analysis provided some evidence of structure.

Introduction

Precise, quantitative data on the structure of fish schools allow the testing of hypotheses concerning the school's adaptive significance or sensory integration. For example, Partridge & Pitcher (1980) have provided convincing evidence that schooling saithe (*Pollachius virens*) use information provided by the lateral line, by examining changes in school structure that followed lateral line section. Partridge & Pitcher (1979) have also effectively refuted Weihs' (1973) model of the hydrodynamic advantages of schooling by comparing its predictions on spacing with actual measurements in the laboratory.

Many workers have attempted to measure the internal structure of fish schools. Studies by Breder (1954, 1959), Hunter (1966) and Van Olst & Hunter (1970) examined the spacing between neighbours in two-dimensional projections (dorsal photographs). However, most schools are three-dimensional entities; two-dimensional measurements will provide interfish distances which are considerably underestimated, and the nearest-neighbour may even be incorrectly identified. Symons (1971b) discusses these difficulties.

Several techniques for three-dimensional mensuration of school structure have also been developed. Apart from the method of Graves (1977), which assumes invariant fish size and uses the size of each

[1] Present address: Marine Mammal Commission, Washington, D.C., U.S.A.

Received 1.2.1980 Accepted 18.8.1980

© Dr. W. Junk b.v. Publishers, The Hague. Printed in The Netherlands. 0378-1909/81/061-007 $01.40.

individual's image on the photograph as a measure of its distance from the camera (i.e. the third dimension), these techniques have been of two major types: stereo and shadow. The stereo method requires obtaining two simultaneous photographs from different angular positions, using either a stereo prism lens (Cullen et al. 1965) or an arrangement of mirrors (Symons 1971a, b, Healey & Prieston 1973, Pitcher 1973, 1975). The shadow method uses the shadows of the fish projected onto the substrate as the second view of the school and has been applied by Dambach (1963), Cullen et al. (1965), Partridge (1980), Partridge & Pitcher (1980), and Partridge et al. (1980).

The above techniques are for the most part highly sophisticated, requiring apparatus which effectively restricts their use to a laboratory situation. There is a great need for a technique which can be used in the field to obtain measurements rapidly of the structures of schools of a variety of species in a range of environmental circumstances. Our group had previously developed a stereo method for measurements of the three-dimensional structure of airborne bird flocks (Major & Dill 1978). This technique, which obtains pairs of photographs simultaneously with two cameras, is here modified and applied to fish schools. Following a description of the technique and a check on its degree of accuracy, we provide some preliminary data on the three-dimensional structure of schools of juvenile coho salmon (*Oncorhynchus kisutch*) swimming in a hatchery raceway.

Methods

Apparatus

Stereo pairs of photographs were taken with two identical motor driven 35-mm cameras (Nikon Photomic F2) with 28 mm lenses. The cameras were mounted side-by-side (32 cm between lens centres) on an aluminum plate (Fig. 1). Each camera 'floated' on a spring-loaded three-point levelling device, and could be independently levelled to an accuracy of 20 s with a level bubble. Care was taken that the two film platens were in the same plane by levelling a plate placed across both of them after removal of the camera backs.

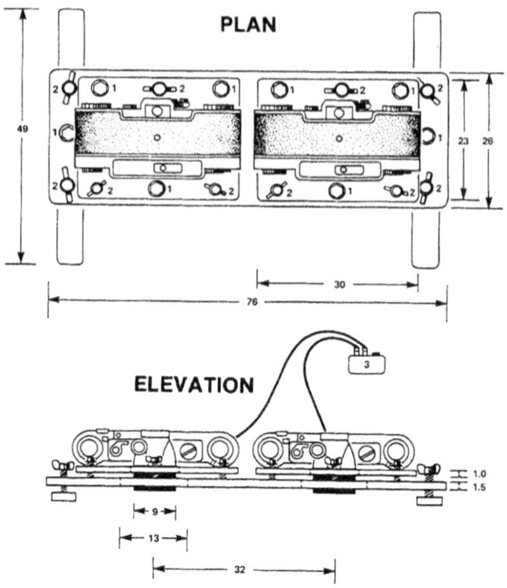

Fig. 1. The camera apparatus used in this study. Two motor-driven 35 mm cameras are mounted on separate levelling platforms, and fired simultaneously (see text for further details).

The entire apparatus was placed on a ladder spanning a hatchery trough, with the cameras pointed downward, their lenses projecting through holes in the aluminum plates. A plexiglass box (open at the top) was suspended below the ladder with its bottom plate a few cm below the water surface. This provided an air-plexiglass-water interface a known distance (462 mm) below the lens faces and eliminated any surface ripple. It was also levelled to an accuracy of 20 s. A black curtain draped over the cameras eliminated excessive surface glare. Water depth (below the plexiglass) was 725 mm.

The cameras were fired simultaneously from a common electronic shutter release box (single frame only; simultaneity checked with a photocell beam through the two camera apertures mounted in tandem). Kodak Plus-X (ASA 125) film was exposed for the minimum time possible given ambient light conditions, and developed according to manufacturer's instructions.

Analysis

The resulting left and right stereo pairs of photographs were analyzed, using a Zeiss-Jena Topocart Analyzer, by Integrated Resources Photography Ltd., Vancouver, B.C. Standard photogrammetric methods were used to determine the three-coordinate position of the tip of the snout of each fish in a school. The procedure incorporated corrections for radial lens distortion characteristics (based on photographs taken of a calibration field in air using the same focal distance and aperture settings as were used in the field) and for refraction (from physical principles and the known geometry of the situation). Numbering the fish on enlarged prints prior to analysis allowed subsequent identification of individuals.

From the three-coordinate data, the following spatial characteristics of each fish's three nearest neighbours were calculated on an IBM 370/148 computer:

$$\text{Distance} = [(\Delta X)^2 + (\Delta Y)^2 + (\Delta Z)^2]^{\frac{1}{2}} \quad \ldots(1)$$

$$\text{Bearing} = \text{TAN}^{-1}\ (\Delta Y / \Delta X) \quad \ldots(2)$$

$$\text{Elevation} = \text{TAN}^{-1}\left[\frac{\Delta Z}{((\Delta X)^2 + (\Delta Y)^2)^{\frac{1}{2}}}\right] \quad \ldots(3)$$

Axis 'X' is the snout-to-tail axis of the fish, but a mean value, taken to be parallel to the walls of the trough, was used in this analysis. Axis 'Y' is perpendicular to the 'X' axis in the same plane, and 'Z' is vertical and perpendicular to the 'X-Y' plane. Mean angle vectors and angular deviations were determined using the methods of Batschelet (1965).

Accuracy

A variety of possible sources of error (discussed in Major & Dill 1978) can reduce the accuracy of measurements obtained using the methods described here. We checked the reliability of our technique by photographing an object whose spatial structure could be independently determined; a lattice structure constructed of laboratory bench hardware. Sixteen black spots (8 mm diameter) were placed at various points on the structure, and the distances between the centres of all possible pairs of spots measured by hand. The lattice was then placed in the water beneath the cameras, photographed, and the three-coordinate position of each spot determined photogrammetrically. Thirteen spots were visible in both photographs, providing 78 'neighbour distances' to compare with those measured directly from the lattice. The average discrepancy in these pairs of measurements was 2.74 mm or 0.67% of the true distance (Fig. 2), and there was no apparent tendency for percentage error to increase with distance (Fig. 3). This level of accuracy was deemed adequate for the present study.

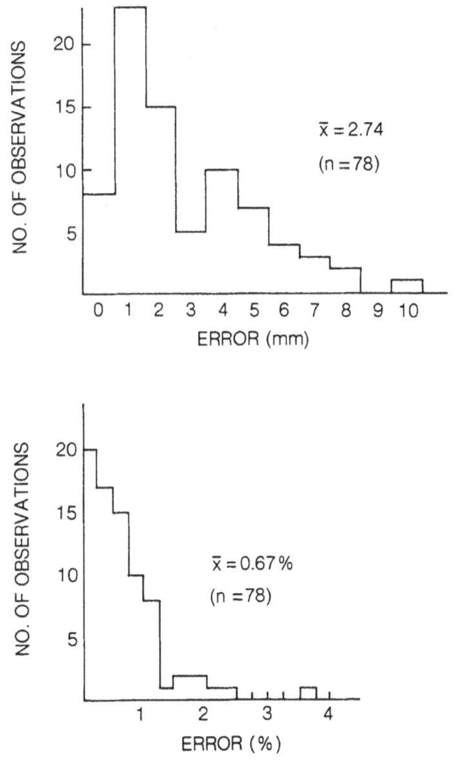

Fig. 2 Frequency distribution of absolute (above) and relative (below) errors in measurements between points on the test lattice.

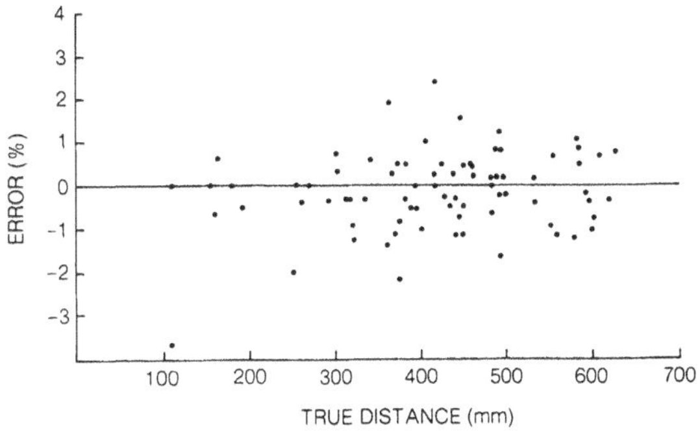

Fig. 3. Percentage measurement error as a function of the distance separating two points on the test lattice.

Application

Stereo pairs of photographs were taken of schools comprising several hundred juvenile coho salmon (*Oncorhynchus kisutch*) swimming in an outdoor trough at the Capilano salmon hatchery, North Vancouver, B.C., on October 5, 1978. The fish averaged 9.6. cm total length (measured from the photographs). The procedures followed were as described above, except that a sheet of opaque white plexiglass (ruled out into 25 × 25 cm squares) was placed on the floor of the trough beneath the cameras in order to enhance contrast. Portions of three separate schools were photographed, and their internal structure analyzed. An example of a stereo pair is shown in Figure 4.

As the salmon reached the edge of the white sheet they appeared to increase their swimming speed, and the cohesiveness of the schools seemed to break down. Cohesiveness was probably not great to start with, since coho salmon are facultative schoolers only. These factors may explain why there is little evidence for any particular structure in the schools photographed. The data are presented more to illustrate the use of the technique than to enhance our understanding of fish school structure.

Frequently, two fish formed a nearest neighbour pair, i.e. they were closer to one another than either was to any other member of the school. Retaining both values in the calculations of mean spacing characteristics can bias the results obtained, in that some values are duplicated in the data set. We have therefore calculated first nearest neighbour distances, bearings, and elevations in two ways: including all fish, and with one member of each nearest neighbour pair excluded. We adopted the convention of discarding data for lead members of all pairs. Mean nearest neighbour distance and mean elevation vector were affected only slightly by this procedure (bearing obviously changes, since only values greater than 90° are eliminated), so it was not repeated for 2nd and 3rd nearest neighbour pairs.

All data for the three schools measured are summarized in Table 1. Not surprisingly inter-fish distances (Fig. 5) increase from 1st through 3rd nearest neighbours. Considering the pooled data, these represent 0.7, 1.0, and 1.2 body lengths, values which correspond well to those determined in other three-dimensional studies (Cullen et al. 1965, Pitcher 1973, Partridge 1980, Partridge et al. 1980).

Mean bearing angles were always close to 90° (i.e. abeam). Note that data for neighbours to the left and right have been pooled in these analyses, as the two distributions did not differ by visual inspection.

Fig. 4. Portions of a stereo pair of photographs of a school of juvenile coho salmon. The disparity in relative positions of the fish (e.g. fish 24 and 25) is an indication of their relative depths in the water.

Table 1. Distances and angles (means and standard or angular deviations) between nearest neighbours in coho salmon schools (ϑ = mean angle vector, S = angular deviation).

School	NN_X	n	*Distance (cm)* $\bar{x}$	± σ	*Bearing (°)* ϑ	± S	*Elevation (°)* ϑ	± S
A	1	44	6.6	2.6	94	45	1	37
	1*	30	7.0	2.8	77	47	2	34
	2	44	9.4	3.5	96	42	−2	32
	3	44	11.4	4.1	90	55	−3	29
B	1	33	6.6	3.1	91	50	3	40
	1*	19	7.0	3.2	53	37	11	38
	2	33	12.1	3.8	85	39	−2	27
	3	33	14.7	4.0	89	33	1	14
C	1	60	6.0	2.1	97	44	−1	35
	1*	41	6.3	2.1	79	44	2	36
	2	60	8.3	2.6	83	42	−1	31
	3	60	10.2	2.7	87	41	−6	28
Pooled	1	137	6.3	2.5	95	46	1	37
	1*	90	6.7	2.6	72	45	3	36
	2	137	9.6	3.2	88	42	−1	31
	3	137	11.7	3.5	89	44	−3	26

*Recalculated for 1st nearest neighbours but with leading members of all nearest neighbour pairs excluded from the analysis.

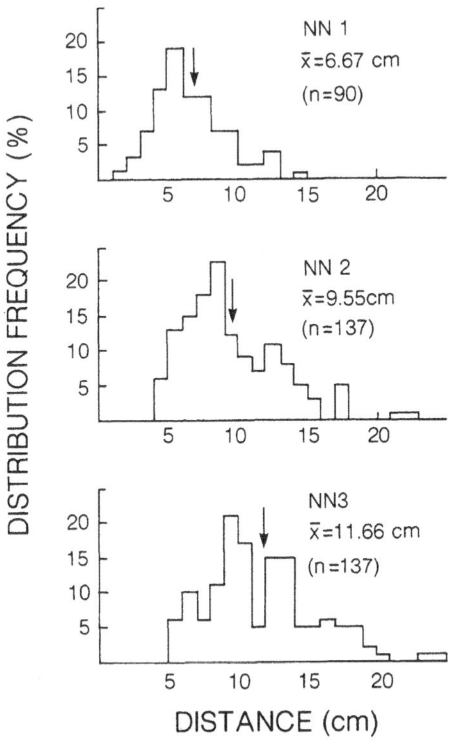

Fig. 5. Frequency distribution of coho first, second, and third nearest neighbour distances (cm). Data for first nearest neighbour exclude leading members of nearest neighbour pairs. Arrows denote mean values.

The frequency distribution of bearing angles was nearly uniform (Fig. 6), except that few nearest neighbours occurred directly in front of or directly behind the reference fish. Mean elevation vectors (Table 1) were always near 0°, i.e. at the same level. Once again this resulted from a nearly uniform frequency distribution (Fig. 7), except that few nearest neighbours occurred directly above or below the reference fish. Similar results have been found for other facultative schoolers (saithe and cod, Partridge et al. 1980), but preferred angular positions for nearest neighbours seem to exist in more obligate schooling species (pilchards, Shaw et al. 1965; minnows, Pitcher 1973, Partridge 1980; herring, Partridge et al. 1980).

If a particular preferred spatial location for nearest neighbours exists in salmon schools, it should reveal itself in a plot of elevation angle against bearing angle. Preferred locations would appear as clusters of data points; none seem to exist in coho salmon, or at least they are not revealed in our relatively small sample (Fig. 8). However, some evidence for a particular structure is provided by the plots of 1st nearest neighbour distance against bearing and elevation angles (Fig. 9). Neighbours are somewhat closer ($p < 0.1$) when diagonally ahead of the reference fish, or when swimming on the same level ($p < 0.05$). Partridge et al. (1980) report

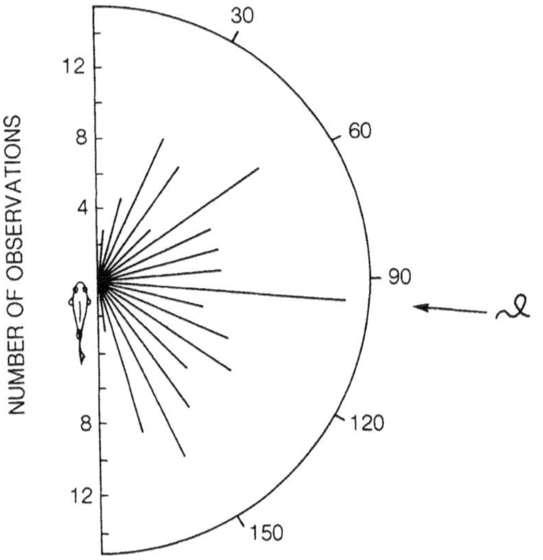

Fig. 6. Frequency distribution of bearing angles for first nearest neighbours in coho schools (n = 137). Arrow denotes mean vector.

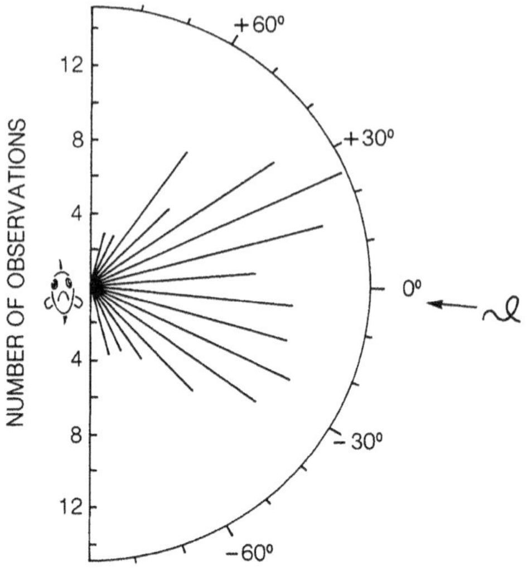

Fig. 7. Frequency distribution of elevation angles for first nearest neighbours in coho schools (n = 137). Arrow denotes mean vector.

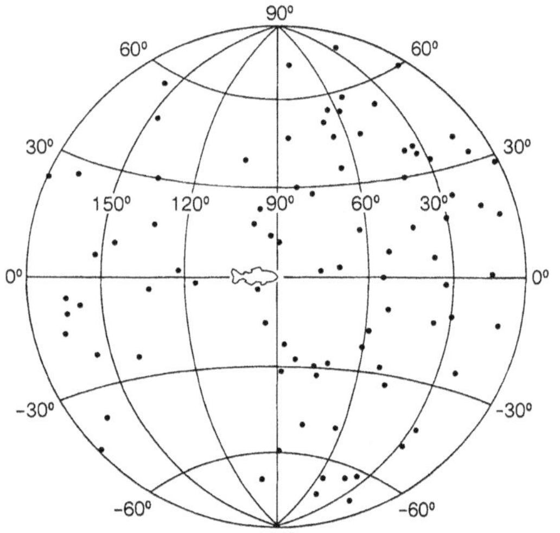

Fig. 8. First nearest neighbour bearing and elevation angles plotted against one another on a hemisphere, for all fish excluding leading members of nearest neighbour pairs (n=90). Each fish is placed at the centre of the hemisphere and the angular location of its neighbour plotted. Each point therefore relates only to the reference fish's position, not to other points.

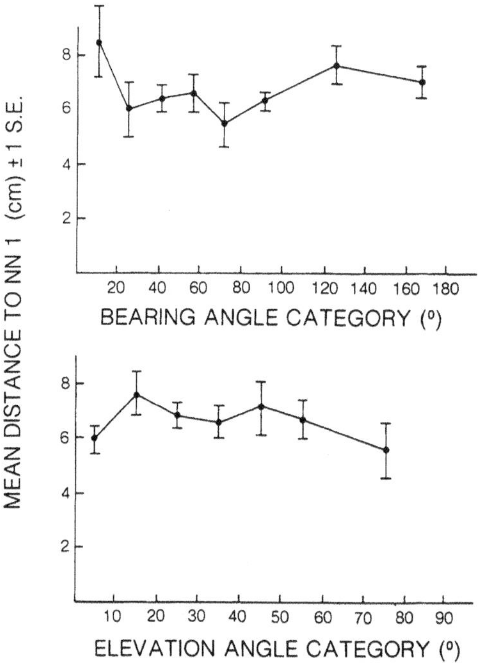

Fig. 9. Mean first nearest neighbour distance (n=90) when neighbours occur in various bearing (above) or elevation (below) intervals. Interval size was varied to keep sample sizes roughly constant. In the elevation plot, similar intervals above and below the reference fish have been combined.

very similar patterns for saithe and cod.

Increasingly it appears that fish schools are not organized in a rigid lattice structure as some have believed (e.g. Breder 1976). Instead the internal structure is in a constant state of flux, as fish change position relative to one another. Any structure is present only in a statistical sense, and can be revealed only by accurate studies in three dimensions, preferably in situations as natural as possible. We believe that our technique, which could fairly readily be adapted for use underwater or in a glass-bottomed boat, provides investigators with a rapid and reliable tool for further studies of this kind.

Acknowledgements

The competent assistance of Jack Russell and his staff at Integrated Resources Photography Ltd. was essential to the success of this study, as was the help of A. Jacques, A. Fraser and particularly R. Ydenberg at various stages in the collection and analysis of the data. F.R. Wick and the SFU machine shop produced the camera mounts to our exacting specifications, and E. Stone gave us permission to work at the Capilano hatchery. B. Partridge kindly provided copies of his unpublished manuscripts and reviewed the paper. The research was funded by an SFU President's Research Grant, and by NSERC Canada grant A6869 to LMD.

References cited

Batschelet, E. 1965. Statistical methods for the analysis of problems in animal orientation and certain biological rhythms. Amer. Inst. Biol. Sci., Washington, D.C. 57 pp.

Breder, C.M., Jr. 1954. Equations descriptive of fish schools and other animal aggregations. Ecology 35: 361–370.

Breder, C.M., Jr. 1959. Studies on social groupings in fishes. Bull. Amer. Mus. Nat. Hist. 117: 393–482.

Breder, C.M., Jr. 1976. Fish schools as operational structures. U.S. Fish. Bull. 74: 471–502.

Cullen, J.M., E. Shaw & H.A. Baldwin. 1965. Methods for measuring the three-dimensional structures of fish schools. Anim. Behav. 13: 534–543.

Dambach, M. 1963. Vergleichende Untersuchungen über das Schwarmverhalten von *Tilapia*-Jungfischen (Cichlidae, Teleostei). Zeit. f. Tierpsychol. 20: 267–296.

Graves, J. 1977. Photographic method for measuring spacing and density within pelagic fish schools at sea. U.S. Fish. Bull. 75: 230–234.

Healey, M.C. & R. Prieston. 1973. The interrelationships among individuals in a fish school. Fish. Res. Board Can., Tech. Rept. 389, 15 pp.

Hunter, J.R. 1966. Procedure for analysis of schooling behavior. J. Fish. Res. Board Can. 23: 547–562.

Major, P.F. & L.M. Dill. 1978. The three-dimensional structure of airborne bird flocks. Behav. Ecol. Sociobiol. 4: 111–122.

Partridge, B.L. 1980. The effect of school size on the structure and dynamics of minnow schools. Anim. Behav. 28: 68–77.

Partridge, B.L. & T.J. Pitcher. 1979. Evidence against a hydrodynamic function for fish schools. Nature 279: 418–419.

Partridge, B. & T.J. Pitcher. 1980. The sensory basis of fish schools: relative roles of lateral line and vision. J. Comp. Physiol. 135: 315–325.

Partridge, B.L., T.J. Pitcher, J.M. Cullen & J. Wilson. 1980. The three-dimensional structure of fish schools. Behav. Ecol. Sociobiol. 6: 277–288.

Pitcher, T.J. 1973. The three-dimensional structure of schools in the minnow, *Phoxinus phoxinus* (L.). Anim. Behav. 21: 673–686.

Pitcher, T.J. 1975. A periscopic method for determining the three-dimensional positions of fish in schools. J. Fish. Res. Board Can. 32: 1533–1538.

Symons, P.E.K. 1971a. Spacing and density in schooling three-spine sticklebacks (*Gasterosteus aculeatus*) and mummichog (*Fundulus heteroclitus*). J. Fish. Res. Board Can. 28: 999–1004.

Symons, P.E.K. 1971b. Estimating distances between fish schooling in an aquarium. J. Fish. Res. Board Can. 28: 1805–1806.

Van Olst, J.C. & J.R. Hunter. 1970. Some aspects of the organization of fish schools. J. Fish. Res. Board Can. 27: 1225–1238.

Weihs, D. 1973. Hydromechanics of fish schooling. Nature 241: 290–291.

Env. Biol. Fish. Vol. 6, No. 1, pp. 15–19, 1981

Computer tools for the analysis of schooling

Michael J. Potel[1] & Richard J. Wassersug[2]

[1]*Department of Biophysics and Theoretical Biology, The University of Chicago, Chicago, Illinois 60637, U.S.A.*
[2]*Department of Anatomy and the Committee on Evolutionary Biology, The University of Chicago, Chicago, Illinois 60637, U.S.A.*

Keywords:
Computer graphics, Interactive computers, Motion analysis, Photogrammetry, Three-dimensional analysis

Synopsis

Quantification of fish school structure is difficult because of: (1) the large amount of positional data that must be recorded, (2) the fact that schools are moving, and (3) the fact that schools are threedimensional. Computer tools for addressing these problems include x,y digitizing devices interfaced to computers and interactive computer graphics systems. General computer algorithms now exist for 3-D reconstruction of schools given any two 2-D views such as photographs.

Our computerized film analyzer, the Galatea system, is described in detail. This system allows for rapid, accurate input of multiple points which can be followed through each frame in a film sequence. Its software packages perform 3-D reconstructions, analyze undulatory motion, and do extensive statistical tests.

Introduction

Our purpose is to provide an overview of certain methods for the quantitative analysis of schools, not to review in detail every method used in the study of schooling. Rather we wish to identify and discuss three broad problems in data acquisition and analysis intrinsic to studies of schools.

The first problem involves the volume of data associated with the study of schools. Given the large number of fish that can form a school, many qualitative studies on schooling have treated the school as a single object and have avoided questions about the interaction of the organisms *within* the school (e.g., studies reviewed in Radakov 1973). However, to get at the internal structure of a school it is necessary to record accurately the positions of all individual organisms comprising the school. Knowledge of the spatial distribution of organisms within a school precedes statistical evaluation of school structure. Thus, the first intrinsic problem with a rigorous examination of schools is the enormous amount of positional information that has to be recorded.

The second intrinsic problem in studying the dynamics of schooling is the analysis of motion. Continuous motion can be reduced to discrete events by a still camera; indeed, early work on school structure involved only still pictures (Cullen et al. 1965, Symons 1971, Pitcher 1973). Meaningful

Received 1.2.1980 Accepted 26.8.1980

© Dr. W. Junk b.v. Publishers, The Hague. Printed in The Netherlands. 0378-1909/81/061-015 $01.00.

analysis of changes in position, however, requires a more dynamic record, such as that provided by video or motion picture cameras. One is reminded that the motion picture camera was invented specifically for the purpose of the analysis of locomotion (Muybridge 1887), yet biological motion analysis has lagged far behind film technology. A scientist with a movie camera can record in a few seconds more raw information on film about the motion of a single object than he can quantify and analyze by hand in months. This is additionally true for studying the dynamics of schooling organisms, because of the large number of objects to follow.

A third intrinsic problem is that of the third spatial dimension. A frequent simplifying assumption is that a school can be studied as a two dimensional problem just as one might study positional relationships of grazing land-based animals. But except in the shallowest of water the spatial arrangement of fish in schools is inherently three-dimensional (Cullen et al. 1965, Pitcher & Partridge 1979). Thus, the proper study of fish schools, or bird flocks (Major & Dill 1978), involves determining the positions of individual organisms in three dimensions.

In the following sections we examine these three issues: multiple data transcription, time-dependent motion analysis, and 3-D analysis. We introduce some of the specific computer tools available, and address the general principles involved so others can find specific solutions appropriate to their studies.

Multiple data transcription

Most laboratory studies on schooling have been limited to schools of very few individuals, although many schools in nature can have hundreds or even thousands of individuals in them. The technical problems with compiling spatial information on individuals in schools has forced most recent workers to seek computer assistance (e.g. Sakamoto et al. 1976, Hurley 1978, Burgess 1979, Pitcher 1979, Partridge 1980, Katz et al. 1981).

The computer can help in measuring positional information from photographic records (i.e. photogrammetry) using techniques from research in fields such as aerial survey and cartography. Relatively inexpensive x,y digitizing devices that interface with computers (Newman & Sproull 1979) transcribe spatial information, such as from a photograph, directly into computer memory.

Computer-interfaced digitizing devices are essential for recording and storing coordinates for large data sets in a practical amount of time. Such devices, however, do not perform pattern recognition (the conversion of pictures to digitized form and subsequent image processing to detect and record features), nor should they. In practice, pattern recognition is computationally expensive, difficult to implement and inflexible in the face of differing images (Duda & Hart 1973, Potel et al. 1980). For fish photographed in clean water, pattern recognition is superfluous, and for analysis of schooling, hand entry of data with a digitizer is faster, less expensive and more accurate.

Digitizer systems with computer feedback offer clear advantages. One such 'interactive' system (Partridge & Cullen 1977), is capable of keeping track of a number of points entered per frame in a film series. This system uses a statistical approach to spot errors in either the order in which points are entered or the distance that a point has moved from frame to frame.

A more directly interactive approach to tracking multiple points is our own Galatea system (Potel et al. 1979, Potel et al. 1980). The system uses a computer graphics display and a computer driven projection cathode ray tube to display on the plotting table wherever a point has been entered with a digitizing pen. The resulting set of projected points, overlaid directly on the source image, provides all the advantages of tracing paper in verifying entries and avoiding omission and duplication of points. The operator need only touch the digitizing pen in illuminated 'control areas' of the plotting surface to count, add or delete points. With such a system, it is possible to record accurately the x,y position of hundreds of points in an hour.

Motion analysis

The Galatea system was specifically designed for motion analysis through the entry of time-dependent data, and has been applied to investigations as disparate as slime mold aggregation (Potel &

MacKay 1979), swimming mechanics of tadpoles (Rubin 1978a, b), and heart function using cine-angiography (Potel et al. 1980). The system incorporates a variable rate stop-motion projector and automatically records every data point as x,y and t, where t is the frame number. As many as 1000 points can be followed from frame to frame (depending upon the total number of frames to be analyzed). As the movie projector is advanced an animated computer graphics representation of the data that have been entered in each frame is superimposed on and synchronized with the film itself. Positions of individual fish can be displayed with different symbols, labels can be added to identify previously entered points and the various moving points can be connected with lines to create animated stick figure representations of the objects under study.

Raw data, recorded as a function of frame number with the computer directing the projection rate can readily be examined with well-developed and widely available techniques for time series analysis (Box & Jenkins 1970, Otnes & Enochsen 1978). We have analyzed undulatory swimming with these procedures (Rubin 1978b). Our programs determine the overall velocities and accelerations of a fish as well as the wave length, velocity and amplitude of the propulsive wave as it travels down the tail. The results are displayed as a variety of ratios (e.g., propellor efficiency), and in units of absolute lengths or body lengths.

3-D analysis

As noted most recently by Pitcher & Partridge (1979), two-dimensional analysis of school geometry can be misleading. Recent advances in our understanding of both the spatial relationships and sensory regulation of schooling (Partridge et al. 1980, Partridge & Pitcher 1980, Pitcher & Partridge 1979) testify to the value of 3-D information.

Fish ethologists have invented for themselves a variety of procedures to determine 3-D positions of fish in schools. These bear names like the stereo, the shadow and the periscope methods (Dambach 1963, Cullen et al. 1965, Symons 1971, Pitcher 1973, 1975, Major & Dill 1978, Dill et al. 1980). Such procedures have been known for a long time in the fields of stereology and photogrammetry (Thompson 1966, Underwood 1970). Ethologists may not have appreciated that all these techniques are special cases of the same general method (Sutherland 1974, Rogers & Adams 1976), which has certain advantages over the specific procedures currently in use. The general case is that 3-D coordinates may be determined from *any* two or more 2-D perspective views, provided these views are known. These perspectives may be front- and side-view photographs, stereo pairs of photographs, a photograph of an object and its view in a mirror, a photograph of an object and its shadow, etc.

The method for inferring 3-D coordinates from 2-D images requires a calibration involving two or more views of a calibration object. The calibration object must contain six or more non-coplanar reference points for which the x,y,z coordinates in the real world are known and for which the x,y coordinates in each 2-D view can be entered. Having calibrated two or more perspective views, the analyst digitizes any 3-D object of interest in both views. The computer will then calculate the actual 3-D position of that object in the coordinate system of the calibration. The only restriction throughout is that all geometric transformations must be linear so that such things as curved film, curved mirrors, 'fish-eye' lenses, etc. cannot be used.

Our computer system is programmed to work with perspective views of this form for multiple 3-D points followed through time (Potel et al. 1980). It has been successfully applied to 3-D reconstruction of blood vessels in the head and the heart from biplanar angiograms (Rubin & Sayre 1978, Sayre et al. 1979, Rubin et al. 1980). In this system, the initial views can be arbitrary, they need not be in fixed relationship as in the stereo method. Perhaps the most important feature of the system is that it does not require persistent fixed reference surfaces, as in the shadow method or methods involving mirrors. The system is unique in providing an arbitrary perspective of the final data on the video screen. For example, a school initially filmed only from the side can be presented on the video screen as it would appear from above, or below, or any other angle.

We believe that the real power of the generalized

3-D analysis will come in describing fish school geometry for fish in natural habitats. A set of stable anchor lines with markers (or fixed landmarks, such as coral heads) and two cameras are the essential equipment necessary to collect data. The cameras are positioned anywhere. All that the investigator needs to know is the 3-D positions of six reference markers on the anchor lines. If the reference markers can be seen in each image, it is even possible to film the scene with non-stationary movie cameras, subsequently calibrating the perspective views in each frame.

Discussion

Not only the quantity but the quality of statistical analyses performed on positional data is enhanced by the computer. Time is inherently sequential, and because of this property, time-dependent data can be subjected to a variety of special techniques that are ideally suited for computer analysis. Many of these time series techniques involve iterative operations to identify lags in phase shifts. Thus, it is possible to examine the lag time in the response of one fish to another (Sakamoto et al. 1976, Partridge 1980) or the lag distance over which one individual interacts with another in terms of number of nearest neighbors, absolute distances or relative body lengths (Katz et al. 1981, Wassersug et al. 1981).

Standard techniques for measuring spatial relationships have been summarized most recently in Southwood (1978). As demonstrated by Stauffer (1979) and Clark & Evans (1979), statistical techniques for 2-D analysis can be extended into 3-D. Also statistical operations that deal with first nearest neighbors can be serially extended to further neighbors once the computer is programmed to find and rank the successive nearest neighbors of individuals in a school. Such techniques have been used by Burgess (1979), Partridge et al. (1980), Partridge (1980), Dill & Major (1980), Katz et al. (1981), and Wassersug et al. (1981) for studies of the geometry of fish and tadpole schools. These techniques are sensitive to structural changes in school geometry that are rarely obvious to the human eye. When they are combined with time series analysis, they provide the best approach to understanding physiological mechanisms underlying school structure.

Although many papers have been published describing techniques for studying schooling (classics include Breder 1954, Cullen et al. 1965, and Hunter 1965), reviews on schooling (most recently Radakov 1973, Shaw 1978, Burgess & Shaw 1979) indicate how little actual quantitative work has been done on school structure. The development of a quantitative approach to the study of schooling behavior has been retarded by the enormous amount of positional information that needs to be analyzed, the problem of tracking objects through time and the problem of tracking objects through three-dimensional space. Recent interactive computer tools, based on techniques that have long been available in other fields, are now being distilled for generalized use in disciplines such as ethology. Although relatively complex at their origin, as implemented in interactive computer graphics they are readily understandable and usable by non-computer specialists.

Dramatic trends are occurring to make small but very powerful computers eminently affordable. The Galatea system was designed in 1973 and constructed over the next several years for approximately US $100,000. The same system today would cost about half that to reproduce and a video version of the system currently under development will cost about US $10,000. As computer prices are driven lower, we expect to see computerized data entry systems become the norm in studies of fish behavior.

Acknowledgements

We thank Larry Katz and Andrew Lum for their collaboration on our schooling studies. Larry Dill and Brian Partridge provided constructive comments on this manuscript. Rashel Paul, Debra Randall, and Andrew Lum helped with the typing. This research has been supported by the Louis Block Fund of the University of Chicago and by NIH grant RR-05367 to MJP; by a Spencer Fund Grant of the University of Chicago and by NSF grant DEB78-23896 to RJW.

References cited

Box, G.E.P. & G.M. Jenkins. 1970. Time series analysis: forecasting and control. Holden-Day, San Francisco, 575 pp.

Breder, C.M., Jr. 1954. Equations descriptive of fish schools and other animal aggregations. Ecology 35: 361–370.
Burgess, J.W. 1979. Measurement of spatial behavior: methodology applied to rhesus monkeys, neon tetras, communal and solitary spiders, cockroaches and gnats in open fields. Behav. & Neural Biol. 26: 132–160.
Burgess, J.W. & E. Shaw. 1979. Development and ecology of fish schooling. Oceanus 22: 11–17.
Clark, P.J. & F.C. Evans. 1979. Generalization of a nearest neighbor measure of dispersion for use in K dimensions. Ecology 60: 316–317.
Cullen, J.M., E. Shaw & H.A. Baldwin. 1965. Methods for measuring the three-dimensional structure of fish schools. Anim. Behav. 13: 534–543.
Dambach, M. 1963. Vergleichende Untersuchungen über das Schwarmverhalten von Tilapia - Jungfischen (Cichlidae, Teleostei). 2. Tierpsychol. 20: 267–296.
Dill, L.M. & R.L. Dunbrack & P.E. Major. 1981. A new stereophotographic technique for analyzing the three-dimensional structure of fish schools. Env. Biol. Fish. 6: 7–13.
Duda, R.O. & P.E. Hart. 1973. Pattern classification and scene analysis. Wiley, N.Y. 482 pp.
Hunter, J.R. 1965. Procedure for analysis of schooling behavior. J. Fish. Res. Board Can. 23: 547–562.
Hurley, A.C. 1978. School structure of the squid *Loligo opalescens*. U.S. Fish. Bull. 76: 433–442.
Katz, L.C., M.J. Potel & R.J. Wassersug. 1981. Structure and mechanisms of schooling in tadpoles of the clawed frog, *Xenopus laevis*. Anim. Behav. 29: 20–33.
Major, P.F. & L.M. Dill. 1978. The three-dimensional structure of airborne bird flocks. Behav. Ecol. Sociobiol. 4: 111–122.
Muybridge, E. 1887. Animal locomotion. University of Pennsylvania. Philadelphia. (Selected reprints In: Animals in Motion, Dover, New York (1957)).
Newman, W.M. & R.F. Sproull. 1979. Principles of interactive computer graphics, 2nd Ed. McGraw-Hill, N.Y. 541 pp.
Otnes, R.K. & L. Enochson. 1978. Applied time series analysis. Vol 1: Basic techniques. Wiley, N.Y. 449 pp.
Partridge, B.L. 1980. The effect of school size on the structure and dynamics of minnow schools. Anim. Behav. 28: 68–77.
Partridge, B.L. & J.M. Cullen. 1977. A low cost interactive coordinate plotter. Behav. Res. Meth. & Instr. 9: 473–479.
Partridge, B.L. & T.J. Pitcher. 1980. The sensory basis of fish schools: relative roles of lateral line and vision. J. Comp. Physiol. 135: 315–325.
Partridge, B.L., T. Pitcher, J.M. Cullen & J. Wilson. 1980. The three-dimensional structure of fish schools. Behav. Ecol. Sociobiol. 6: 277–288.
Pitcher, T.J. 1973. The 3-dimensional structure of schools in the minnow (*Phoxinus phoxinus*). Anim. Behav. 21: 673–686.
Pitcher, T.J. 1975. A periscopic method for determining the three-dimensional positions of fish in schools. J. Fish. Res. Board Can. 32: 1533–1538.
Pitcher, T. 1979. Sensory information and the organization of behavior in a shoaling cyprinid fish. Anim. Behav. 27: 126–149.
Pitcher, T.J. & B.L. Partridge. 1979. Fish school density and volume. Mar. Biol. 54: 383–394.
Potel, M.J. & S.A. MacKay. 1979. Preaggregative cell motion in *Dictyostelium*. J. Cell. Sci. 36: 281–309.
Potel, M.J., R.E. Sayre & S.A. MacKay. 1980. Graphics input tools for interactive motion analysis. Comput. Graphics & Image Processing 14: 1–23.
Potel, M.J., R.E. Sayre & A. Robertson. 1979. A system for interactive film analysis. Comput. Biol. Med. 9: 237–256.
Radakov, D.V. 1973. Schooling in the ecology of fish. John Wiley & Sons, New York, New York. 173 pp.
Rogers, D.J. & J.A. Adams. 1976. Mathematical elements for computer graphics. McGraw-Hill, N.Y. 239 pp.
Rubin, J.M., N.J. Patronas, E.E. Duda, R.E. Sayre & M.J. Potel. 1980. Clinical applications of combined cerebral angiograms and brain CT scans. Amer. J. Neuroradiology 1: 83–87.
Rubin, J.M. & R.E. Sayre. 1978. A computer-aided technique for overlaying cerebral angiograms onto computed tomograms. Investigative Radiology 13: 362–367.
Rubin, J. 1978a. A comparative study of tadpole locomotion. Amer. Zool. 17: 973.
Rubin, J. 1978b. Computer-aided film analysis of tadpole swimming. M. Sc. Thesis, Dept. of Anatomy, Univ. of Chicago. 39 pp.
Sakamoto, W., T. Inagaki & T. Kuroki. 1976. Studies on the schooling behavior of fish – IV some discussions concerning continuity of swimming speed and response time of individuals. Bull. Jap. Soc. Sci. Fisheries 42: 1083–1091.
Sayre, R.E., J.M. Rubin, E.E. Duda & N.J. Patronas. 1979. Quantitative three-dimensional angiograms: applications, including augmentation of computed tomograms. pp. 95–102. Proc. IEEE 6th Conf. on Comput. Applic. in Radiology, Newport Beach, CA.
Shaw, E. 1978. Schooling fishes. Amer. Sci. 66: 66–125.
Southwood, T.R.E. 1978. Ecological methods. Chapman and Hall, London, England. 524 pp.
Stauffer, H.B. 1979. A generalization of Pielou's index of nonrandomness. J. Theor. Biol. 77: 19–25.
Sutherland, I.E. 1974. Three-dimensional data input by tablet. Proc. IEEE 62: 453–461.
Symons, P.E.K. 1971. Spacing and density in schooling threespine sticklebacks (*Gasterosteus aculeatus*) and mummichog (*Fundulus heteroclitus*). J. Fish. Res. Board Can. 28: 999–1004.
Thompson, M.M. (ed.) 1966. Manual of photogrammetry, 3rd. Ed. American Society of Photogrammetry, Washington, D.C. 1199 pp.
Underwood, E.E. 1970. Quantitative Stereology. Addison-Wesley, Reading, M.A. 274 pp.
Wassersug, R., A. Lum & M. Potel. 1981. An analysis of school structure for tadpoles (Anura: Amphibia). Behav. Ecol. Sociobiol. (submitted).

Env. Biol. Fish. Vol. 6, No. 1, pp. 21–24, 1981

Facultative schooling behavior in the spottail shiner (*Notropis hudsonius*): possible costs and benefits

Benoni H. Seghers

Department of Zoology, University of Western Ontario, London, Ontario N6A 5B7, Canada

Keywords:
Evolution of sociality, Antipredator behavior, Vigilance, Feeding behavior, *Perca flavescens*

Synopsis

If an individual is less susceptible to predation in a group it should be able to reduce its vigilance for predators and perhaps also its tendency to flee from predators, thereby gaining time and energy for feeding and other activities. These predictions were examined in the field (Veronica Lake, northwestern Ontario) using a facultative schooling species, the spottail shiner (*Notropis hudsonius*). Schools of various sizes 'attacked' with a fish predator model had a significantly lower reaction distance to the predator than solitary shiners. Solitary shiners must therefore invest more time and energy in defense but may gain a feeding advantage because their visual field does not overlap with conspecific competitors.

Introduction

Theoretical models in sociobiology have emphasized the role of food resources and predator defense as major ultimate determinants of sociality (Alexander 1974, Brown 1975, Wilson 1975, Bertram 1978, Rubenstein 1978). If an individual is less susceptible to predation in a group it should be able to reduce its vigilance for predators and perhaps also its tendency to flee from predators thereby gaining time and energy for feeding and other essential activities. Several studies of gregariousness in birds (Lazarus 1978, 1979, Lazarus & Inglis 1978, Powell 1974, Siegfried & Underhill 1975, Smith 1977) and mammals (Hamilton et al. 1978, Berger 1978, Hoogland 1979) have examined these predictions but it has not been established if fish schools conform to this pattern. In this paper I present field data showing that spottail shiners (*Notropis hudsonius*) in schools are less responsive than solitary shiners to a potential predator.

Received 2.1.1980 *Accepted 24.8.1980*

Field observations

I studied the behavior of juvenile shiners (< 2 mo old) in a shallow sandy bay of Veronica Lake, a 120 ha oligotrophic lake 50 km southeast of Kenora, Ontario, Canada. The shiners have a relatively simple time-energy budget at this early life history interval: they must feed on zooplankton, grow quickly before the onset of the five-month period of ice cover, and escape the attacks of juvenile and adult yellow perch (*Perca flavescens*), the only major piscivores present during the summer growth period.

Most of the juvenile shiners occur within 2 m of shore in water less than 25 cm deep. The shiners share this habitat with the juvenile perch who spend considerable time and effort chasing the shiners,

© Dr. W. Junk b.v. Publishers, The Hague. Printed in The Netherlands. 0378-1909/81/061-021 $00.80.

thus forcing the shiners to interrupt their feeding, alter their swimming path, and often make vigorous escape movements. The size relationship between juvenile shiners and juvenile perch is illustrated by samples taken with a fine mesh seine in mid-July 1978 (Fig. 1). Most of the predator-prey interactions occur between the larger members of the juvenile perch stock and the smaller members of the shiner stock. Older perch (1+ and 2+ year class) appear infrequently in the shallow water as they dart in from deeper water to attack both the shiners and juvenile perch. The antipredator behavior of the shiners resembles the description given by Nursall (1973) for adult spottail shiners in Alberta, Canada.

The shiners may occur singly or in schools, i.e. they are facultative schoolers (Shaw 1978). Schools consist of two or more fish of similar size swimming together in a coordinated (usually polarized) fashion, 1 to 6 body lengths apart. Single fish are not affiliated with a school and are usually farther than 50 cm from the nearest shiner. They may, however, eventually form or join a school. Few schools maintain their integrity for more than a minute (Evered & Seghers, in preparation) supporting Radakov's (1973) generalization that the composition of schools of planktivorous fish is highly unstable.

Size of shiner schools

To determine the number of fish in a school I walked slowly along a 145 m beach and recorded the frequency of single fish and fish in various school sizes. To obtain accurate estimates for schools larger than five fish, repeated counts were made using a mechanical tally counter. Observations were made in the late afternoon (1620 - 1930 h); water temperature varied from 19.5° C to 22.5° C and illuminance (measured over the water surface with a light meter) ranged from 11,000 to 60,000 1×.

School size varied from 2 to 49 fish but in 39.8% (195/490) of the observations or 12.3% (195/1587) of the total number of fish counted, the shiners were not schooled (Fig. 2). Almost 50% of the shiners were distributed as solitary fish and schools of less than five fish.

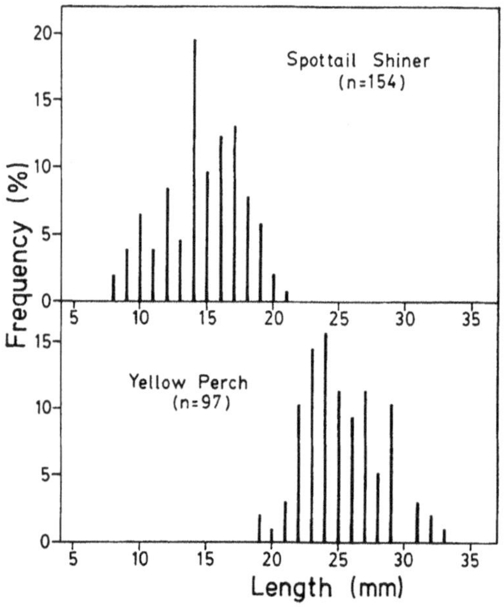

Fig. 1. Length (total) frequency histogram of juvenile spottail shiners (prey) and juvenile yellow perch (predators) co-occurring in the littoral zone of Veronica Lake, northwestern Ontario. Samples were taken in mid-July, 1978.

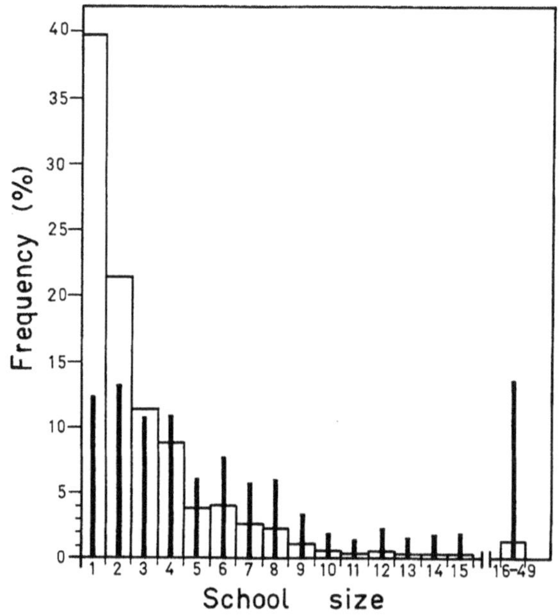

Fig. 2. Distribution of school sizes and individual membership in schools of juvenile spottail shiners, Veronica Lake, mid-July, 1978. Histograms indicate frequency of schools of a particular size (490 observations), whereas black bars indicate the percent of 1587 individuals that occurred in schools of a particular size.

Response to simulated predator attack

I measured the response of the shiners to a potential predator by 'attacking' a single fish or an entire school with a plastic 10 cm fishing lure (hooks removed) resembling a pike (overall green in color and countershaded) mounted on the end of a 1.8 m glass rod. The attack began 1 m from the fish at a constant approach velocity of 20 cm s^{-1} (controlled manually). I recorded to the nearest 10 cm the distance at which the individual fish or school fled from the predator model ('reaction distance'). Because it was not possible to perceive individual differences in reaction distance among school members, a single distance was recorded for the entire school. Control tests were conducted in an identical manner without a lure on the glass rod. This resulted in reaction distances of 0 – 5 cm indicating that the fish were reacting to the predator model and not my movements.

Results of 176 tests revealed that shiners in schools had a significantly shorter reaction distance than solitary shiners ($p < 0.001$, t-test); however, there was no difference among the various school sizes (Fig. 3).

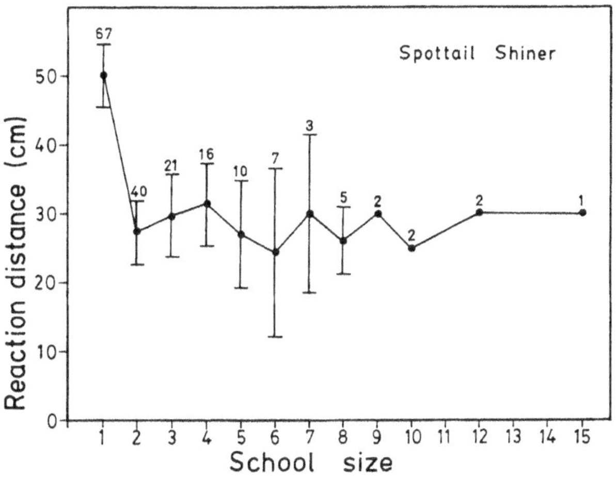

Fig. 3. The relationship between school size of spottail shiners and reaction distance to a predator model. Tests were conducted on calm, sunny days (approximately 0900 - 1800 h) in mid-July, 1978. The number of replicate tests is shown above each mean; vertical lines are ± 2 SE.

Discussion

The relative probability that an individual will be captured, once detected, should be approximately 1/N where N is the school size. It is therefore surprising that the presence of only a single companion had the same effect on the escape threshold as many companions. Theoretically individuals in schools larger than two could have 'afforded' to relax their reaction distances more. The escape threshold of the juvenile shiner therefore seems to have only two basic settings, high for individuals and low for schools.

Although the reaction distance for escape did not change as a function of school size, it is possible that the predator was detected sooner or at a greater distance in larger schools (cf. Nikolsky 1955, cited in Radakov 1973) as has been demonstrated for other vertebrate social groups (Powell 1974, Siegfried & Underhill 1975, Carl 1971, Hoogland & Sherman 1976, Kenward 1978, Hoogland 1979). This phenomenon has been termed the 'early warning function' of social groups (Lazarus 1979). Under the present field conditions it was not possible to discern subtle changes in behavior such as eye movements or orienting reactions. Laboratory experiments are in progress to examine this (Abrahams & Seghers, in preparation).

Why do solitary fish occur so frequently? This is puzzling because recent field work in Veronica Lake (Seghers, in preparation) has revealed that the perch attack solitary shiners preferentially. Also, there is a growing body of evidence confirming that single fish are indeed more vulnerable to predators than schooled fish (Radakov 1973, Neill & Cullen 1974, Major 1978, McKaye et al. 1979). Single fish have a large 'domain of danger' (Hamilton 1971) and when viewed against the visual background of a school they might appear spatially odd to a predator (Milinski 1977) and hence be at a greater risk. Moreover, they cannot benefit from the 'confusion effect' of schools (Hobson 1978, Moss 1979).

Although the shiners presumably could reduce the time and energy allocated to antipredator behavior by associating permanently in schools, there is a potential benefit to solitary fish that may partially offset the increased cost of defense. Eggers (1976) has shown mathematically that schooling is disadvantageous for planktivores because the visual field of each individual overlaps partially with that of others in the school. This may result in competition for zooplankton. If this hypothesis is correct, shiners may alternate between a schooled and solitary strategy as a function of hunger. This is supported by the general observation that feeding fish form less cohesive schools (Keenleyside 1955, Hunter 1966). Perhaps the high frequency of very small schools observed in this study reflects the best compromise between avoiding intraspecific competition for food and predation by perch.

Acknowledgements

I thank B.M. Horn and D.L. Kramer for comments on this study and M.A. Evered, G.S. Helfman and L.E. Roth for critically reading the manuscript. Financial support was from NSERC of Canada grant A9894.

References cited

Alexander, R.D. 1974. The evolution of social behavior. Ann. Rev. Ecol. Syst. 5: 325–383.

Berger, J. 1978. Group size, foraging, and antipredator ploys: an analysis of bighorn sheep decisions. Behav. Ecol. Sociobiol. 4: 91–99.

Bertram, B.C.R. 1978. Living in groups: predators and prey. pp. 64–96. In: J.R. Krebs & N.B. Davies (ed.) Behavioural Ecology: An Evolutionary Approach, Sinauer, Sunderland.

Brown, J.L. 1975. The evolution of behavior. W.W. Norton Co., New York. 761 pp.

Carl, E.A. 1971. Population control in arctic ground squirrels. Ecology 52: 395–413.

Eggers, D.M. 1976. Theoretical effect of schooling by planktivorous fish predators on rate of prey consumption. J. Fish. Res. Board Can. 33: 1964–1971.

Hamilton, W.D. 1971. Geometry for the selfish herd. J. theor. Biol. 31: 295–311.

Hamilton, W.J., III, R.E. Buskirk & W.H. Buskirk. 1978. Omnivory and utilization of food resources by chacma baboons, *Papio ursinus*. Amer. Natur. 112: 911–924.

Hobson, E.S. 1978. Aggregating as a defense against predators in aquatic and terrestrial environments. pp. 219–234. In: E.S. Reese & F.J. Lighter (ed.) Contrasts in Behavior, John Wiley & Sons, New York.

Hoogland, J.L. 1979. The effect of colony size on individual alertness of prairie dogs (Sciuridae: *Cynomys* spp.). Anim. Behav. 27: 394–407.

Hoogland, J.L. & P.W. Sherman. 1976. Advantages and disadvantages of bank swallow (*Riparia riparia*) coloniality. Ecol. Monogr. 46: 33–58.

Hunter, J.R. 1966. Procedure for analysis of schooling behavior. J. Fish. Res. Board Can. 23: 547–562.

Keenleyside, M.H.A. 1955. Some aspects of the schooling behaviour of fish. Behaviour 8: 183–248.

Kenward, R.E. 1978. Hawks and doves: factors affecting success and selection in goshawk attacks on woodpigeons. J. Anim. Ecol. 47: 449–460.

Lazarus, J. 1978. Vigilance, flock size and domain of danger size in the white-fronted goose. Wildfowl 29: 135–145.

Lazarus, J. 1979. The early warning function of flocking in birds: an experimental study with captive quelea. Anim. Behav. 27: 855–865.

Lazarus, J. & I.R. Inglis. 1978. The breeding behaviour of the pink-footed goose: parental care and vigilant behaviour during the fledging period. Behaviour 65: 62–88.

Major, P.F. 1978. Predator-prey interactions in two schooling fishes, *Caranx ignobilis* and *Stolephorus purpureus*. Anim. Behav. 26: 760–777.

McKaye, K.R., D.J. Weiland & T.M. Lim. 1979. Comments on the breeding biology of *Gobiomorus dormitor* (Osteichthyes: Eleotridae) and the advantage of schooling behavior to its fry. Copeia 1979: 542–544.

Milinski, M. 1977. Experiments on the selection by predators against spatial oddity of their prey. Z. Tierpsychol. 43: 311–325.

Moss, R.E. 1979. Miller's Law and the clustering of prey. Unpublished manuscript. 26 pp.

Neill, S.R. St.J. & J.M. Cullen. 1974. Experiments on whether schooling by their prey affects the hunting behaviour of cephalopods and fish predators. J. Zool (Lond.) 172: 549–569.

Nursall, J.R. 1973. Some behavioral interactions of spottail shiners (*Notropis hudsonius*), yellow perch (*Perca flavescens*), and northern pike (*Esox lucius*). J. Fish. Res. Board Can. 30: 1161–1178.

Powell, G.V.N. 1974. Experimental analysis of the social value of flocking by starlings (*Sturnus vulgaris*) in relation to predation and foraging. Anim. Behav. 22: 501–505.

Radakov, D.V. 1973. Schooling in the ecology of fish. John Wiley & Sons, New York. 173 pp.

Rubenstein, D.I. 1978. On predation, competition, and the advantages of group living. pp. 205–231. In: P.P.G. Bateson & P.H. Klopfer (ed.) Perspectives in Ethology, Plenum, New York.

Shaw, E. 1978. Schooling fishes. Amer. Sci. 66: 166–175.

Siegfried, W.R. & L.G. Underhill. 1975. Flocking as an antipredator strategy in doves. Anim. Behav. 23: 504–508.

Smith, J.N.M. 1977. Feeding rates, search paths, and surveillance for predators in great-tailed grackle flocks. Can. J. Zool. 55: 891–898.

Wilson, E.O. 1975. Sociobiology: the new synthesis. Harvard Univ. Press, Cambridge. 697 pp.

Env. Biol. Fish. Vol. 6, No. 1, pp. 25–29, 1981

Effects of alarm substance on schooling in the common shiner (*Notropis cornutus*, Cyprinidae)

Eva J. Heczko & Benoni H. Seghers

Department of Zoology, University of Western Ontario, London, Ontario N6A 5B7, Canada

Keywords:
Schreckstoff, Fright reaction, Pheromone, Antipredator mechanisms, Cover-seeking, Fish

Synopsis

The behavioral responses of common shiners (*Notropis cornutus*) to 'Schreckstoff' (alarm pheromone produced in the epidermis) were quantified in laboratory experiments. Schreckstoff increased cohesion and polarization, and decreased the variability in overall school dimensions. There was also a positive correlation between size of fish and their distance to the center of the school. Furthermore, the tendency to seek cover increased in the presence of Schreckstoff. The greater organization in school structure appears to be an adaptive response to aquatic predators, whereas increased cover seeking may be an adaptive response to aerial predators.

Introduction

'Schreckstoff' is an alarm pheromone (von Frisch 1938, in Pfeiffer 1962, Smith 1977) released from club cells in the epidermis (Pfeiffer & Lemke 1973) when damaged (e.g. when a fish is captured by a predator). Other fish will respond to this water-borne pheromone with a 'Schreckreaktion' (fright reaction), which may consist of cover seeking, closer crowding, rapid swimming or immobility (von Frisch 1938, 1941, in Pfeiffer 1962, Verheijen 1962a, 1963). Detection of the pheromone is only olfactory (Pfeiffer 1963b), but the fright reaction may be visually transmitted to other fish (Verheijen 1956). The initial reaction involves increased respiration rate (Pfeiffer 1962) which may be accompanied by bradycardia (Pfeiffer & Lamour 1976) and increased visual alertness (Pfeiffer & Riegelbauer 1978).

Schreckstoff and the Schreckreaktion appear to be confined to the Ostariophysi (Pfeiffer 1977). The exact molecular structure of the alarm substance is not yet known (Pfeiffer 1975, 1978), but it seems similar in all species, since it is effective interspecifically with the strongest reaction between closely related species (Pfeiffer 1963a, Smith 1977). Most authors usually ascribe an antipredator function to this pheromone, through inhibition of cannibalism (Verheijen 1962b, Pfeiffer 1963a, Verheijen & Reuter 1969, Marusov 1976), repulsion of predators (Williams 1964), or by warning others in a school (Pfeiffer 1962).

This study tested the third hypothesis, that it functions as an advance alert mechanism. We attempted to quantify the structural changes in schools of shiners upon addition of Schreckstoff,

Received 30.1.1980 Accepted 31.7.1980

© Dr. W. Junk b.v. Publishers, The Hague. Printed in The Netherlands. 0378-1909/81/061-025 $01.00.

following the suggestion for more objective quantification of the alarm response (Aoki & Kuroki 1975).

Materials and methods

Pretreatment and preparation of Schreckstoff

Common shiners were seined from Medway Creek, a tributary of the Thames River, near London, Ontario. They were held in 200 l tanks at 19 – 22° C, on a 10:14 h light: dark photoperiod, and fed dry fish food (Tetramin) to satiation each morning.

Schreckstoff was freshly prepared before each experiment by decapitating a shiner (6 cm total length), eviscerating the body, removing the scales (to release Schreckstoff), and cutting the skin 10 times (to a depth of 1 mm) on both sides. After soaking the body in 75 ml of aged water for 5 min the solution was filtered (0.1 cm pore size). The filtrate was a potent stimulus for the fright reaction.

Effect of Schreckstoff on school structure

Although we observed shiners in schools of up to several hundred in Medway Creek, we used only six fish in each test school so behavior could be monitored for individuals. A white, circular pool (1.82 m diameter) was filled with aged water to a depth of 10 cm, well within the range observed in Medway Creek, but shallow enough to make the school virtually two-dimensional (Williams 1964, Van Olst & Hunter 1970). A clear plastic (polyethylene) liner in the pool was discarded with the water after each use. An overhead 400 watt incandescent bulb provided illumination.

Six fish of equal length (± 0.3 cm) were placed into the pool and left undisturbed for 1 hour. Then either 2 l of aged water (control) or 75 ml of Schreckstoff extract followed by 2 l aged water was poured in through a hose near the edge of the pool. Immediately after the introduction of this water, five photographs were taken, without flash at 1 min intervals with black and white film (ASA 400). This was repeated for 20 different schools (10 control, 10 Schreckstoff). A minimum of three photographs from each school were used for analyses.

Each negative was projected on a 18 × 24 cm grid divided into 1 cm squares. The X-Y coordinates of each fish's snout were plotted, and the center of the school determined by calculating the mean X-Y coordinates. The mean of the distances of individual fish from the center of the school was used as a measure of overall cohesion (Thines & Vandenbussche 1966). A decrease in mean distance to center of the school indicated an increase in overall cohesion. A measure of school polarization was obtained by determining the heading of each fish with a protractor, and calculating the mean heading for the school. The deviation of each fish from the mean heading was calculated, and a mean obtained for the school (Van Olst & Hunter 1970).

Effect of Schreckstoff on cover-seeking behavior

A white, circular pool (1.21 m diameter) was lined with a polyethylene plastic sheet and filled with aged water to a depth of 15 cm. A clump of submerged vegetation approximately 15 cm in diameter was placed near the edge. A second pool (control) was left with only water and vegetation, while Schreckstoff extract was poured into the experimental pool and mixed thoroughly. Six common shiners were then placed into either pool and recordings of their behavior begun immediately. The number of fish in vegetation over a 1 min period was recorded (from behind a blind) with one channel of an event recorder (Esterline-Angus) for each fish, every 5 min over a 1 hour period.

Results and discussion

Schreckstoff caused a significant increase (Mann-Whitney U-test, $p<0.05$) in cohesion among shiner schools (Fig. 1). This result concurs with the increased cohesion among schools of *Rasbora heteromorpha* in response to Schreckstoff (Thines & Vandenbussche 1966). Significantly less variation (F-test, $p<0.01$) was found in the mean distance of school center in the presence of Schreckstoff extract (Fig. 1). In addition, a significant positive correlation ($p<0.05$) was found between size of fish and mean distance to center (Fig. 2). It is to be expected that larger fish have a greater mean distance to school center for physical reasons. This relationship was not found for control groups, so the treated

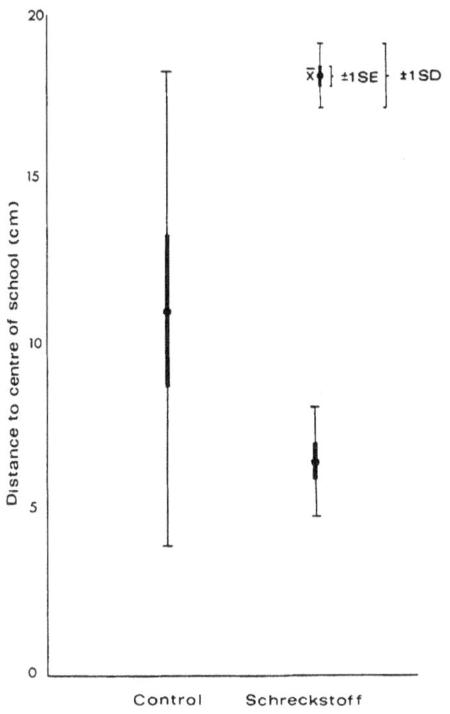

Fig. 1. The effect of Schreckstoff on school cohesion (as measured by the mean distance to the centre of school, Control N=10, Schreckstoff N=10).

schools not only exhibited greater cohesion but also seemed to pack together as closely as their size would permit. Furthermore, Schreckstoff treated schools were significantly more polarized than controls (Fig. 3).

The changes in the Schreckstoff treated schools mean they were more uniform or structured (Shaw 1978, p. 66). This decreases the probability of spatially odd individuals, such individuals being more susceptible to predation (Mueller 1971, Milinski 1977). The greater organization of the school should also enhance communication through a variety of sensory channels.

This enhanced structural organization supports the theory of schooling as an antipredator mechanism (Hobson 1968). The confusion effect from a number of closely-spaced fish suggests that predators are unable to fixate on several moving targets simultaneously, but must isolate a single fish. Rock bass, *Ambloplites rupestris*, prey on schools of common shiners by striking at the school to scatter the individuals, and then feeding on isolated prey (Heczko 1980). Increased school cohesion has been reported in response to predators in sticklebacks, *Gasterosteus aculeatus*, and guppies, *Poecilia reticulata* (Keenleyside 1955, Seghers 1974).

Shiners exposed to Schreckstoff extract spent more time hidden in vegetation than did controls (Fig. 4). The initial response of shiners provided with vegetation is to seek cover. The presence of Schreckstoff causes the fish to remain hidden and not begin exploratory movements as they would normally. This minimizes movement, and so may further decrease detection by predators (Edmunds 1974). But when does the fright reaction consist of cover seeking and when does it produce increased structural organization of the school?

If schooling is simply a form of cover seeking (Williams 1964), the answer is obvious. In the absence of other kinds of cover, fish take cover among themselves (i.e. increased school cohesion). Another possibility is that cover seeking may occur mainly in shallow water. Remains of shiners have been found in the stomachs of a number of piscivorous birds which feed in shallow water (Salyer & Lagler 1940, 1946, Kushlan 1978). Moving a dark ob-

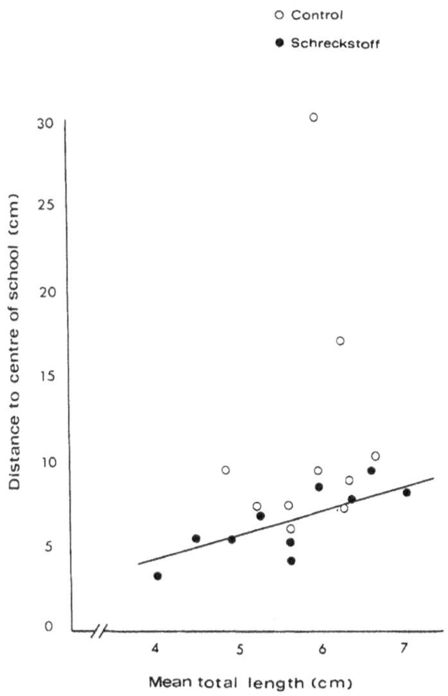

Fig. 2. The relationship of distance to centre of school with mean size of shiner. Schools exposed to Schreckstoff exhibit a significant positive correlation ($r=0.724$, $p<0.05$, regression line of best fit).

ject over an experimental pool, to simulate an aerial predator, always elicited cover-seeking behavior in the shiners. But shiners never entered vegetation when a rock bass was present in the pool (Heczko 1980). Thus, cover seeking may be the primary defense mechanism (Edmunds 1974), and increased cohesion the secondary mechanism (avoidance of being captured once detected).

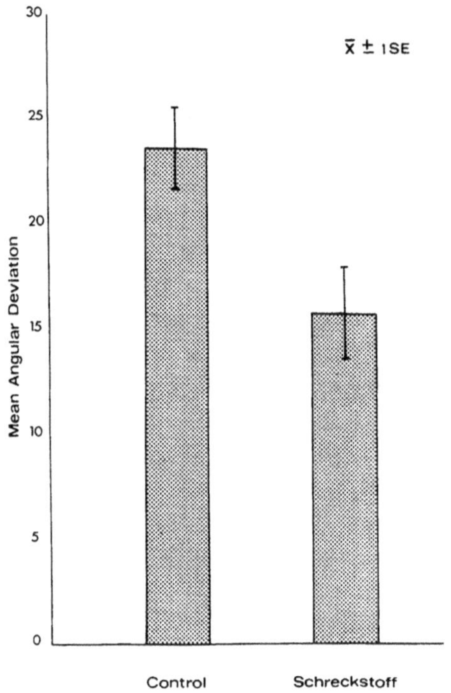

Fig. 3. The effect of Schreckstoff on the degree of polarization of a school of six common shiners (measured as mean angular deviation, Van Olst & Hunter 1970).

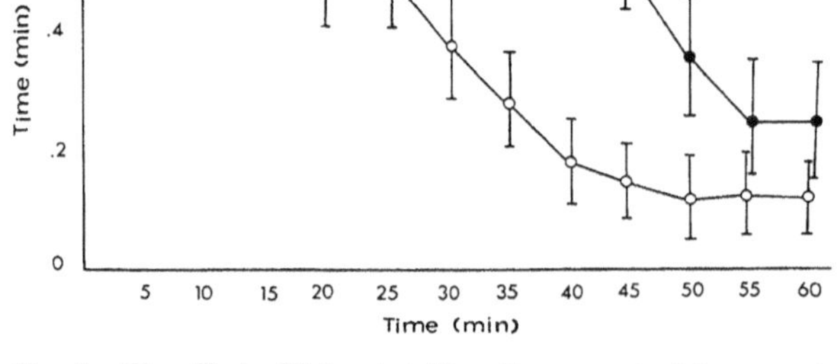

Fig. 4. The effect of Schreckstoff on the amount of time spent in vegetation by a school of six common shiners (Control N = 10, Schreckstoff N = 12).

Acknowledgements

This paper is based on an M.Sc. thesis at The University of Western Ontario, supported by a N.S.E.R.C. research grant to B.H. Seghers.

Many thanks to Brian Bietz, Hélène Dupuis and Graham Sudbury for help in collection of the fish, John Englert for encouragement and assistance during the course of the study, Roger Green for statistical guidance and Jean-Guy Godin for critical reading of the initial draft.

References cited

Aoki, I. & T. Kuroki. 1975. Alarm reaction of three Japanese cyprinid fishes, *Tribolodon hakonensis*, *Gnathopogon elongatus elongatus*, *Rhodeus ocellatus ocellatus*. Bull. Jap. Soc. Sci. Fish. 41: 507–513.

Edmunds, M. 1974. Defence in animals. Longman, Harlow. 357 pp.

Frish, K. von. 1938. Zur Psychologie des Fish-Schwarmes. Die Naturwissenschaften 26: 601–606.

Frisch, K. von. 1941. Über einen Schreckstoff der Fischhaut und seine biologische Bedeutung. Z. vergl. Physiol. 29: 46–145.

Heczko, E. 1980. The effect of Schreckstoff on the behavior of the common shiner (*Notropis cornutus*, Cyprinidae). M.Sc. Thesis, University of Western Ontario, London. 67 pp.

Hobson, E.S. 1968. Predatory behavior of some inshore fishes in the Gulf of California. U.S. Bureau of Sport Fisheries and Wildlife. Research Report 73.

Keenleyside, M.H.A. 1955. Some aspects of the schooling behaviour of fish. Behaviour 8: 183–248.

Kushlan, J.A. 1978. Feeding ecology of wading birds. pp 249–297. *In*: A. Sprunt, J.C. Ogden, S. Winckler (ed.) Wading birds, Research Report # 7 of the National Audubon Society.

Marusov, Ye. A. 1976. Fright reaction in the predatory cyprinid *Opsariichthys uncirostris amunsis*. J. Ich. 16: 685–688.

Milinski, M. 1977. Experiments on the selection by predators against spatial oddity of their prey. Z. Tierpsychol. 43: 311–325.

Mueller, H.C. 1971. Oddity and specific search image more important than conspicuousness in prey selection. Nature 233: 345–346.

Pfeiffer, W. 1962. The fright reaction of fish. Biol. Rev. 37: 495–511.

Pfeiffer, W. 1963a. The fright reaction in North American fish. Can. J. Zool. 41: 69–77.

Pfeiffer, W. 1963b. Alarm substances. Experientia 19: 113–123.

Pfeiffer, W. 1975. Fluorescent pterins from the skin of Cypriniformes (Pisces) and their relation to the alarm pheromone. Rev. Suisse Zool. 82: 705–711.

Pfeiffer, W. 1977. The distribution of fright reaction and alarm substance cells in fishes. Copeia 1977: 653–665.

Pfeiffer, W. 1978. Heterocyclic compounds as releasers of the fright reaction in the giant danio, *Danio malabaricus* (Jerdon) (Cyprinidae, Ostariophysi, Pisces). J. Chem. Ecol. 4: 665–673.
Pfeiffer, W. & D. Lamour. 1976. The action of the alarm substance on the heart frequency of *Phoxinus phoxinus* (Cyprinidae, Ostariophysi, Pisces). Rev. Suisse Zool. 83: 861–873.
Pfeiffer, W. & J. Lemke. 1973. Untersuchungen zur Isolierung und Identifizierung des Schreckstoffes aus der Haut der Elritze *Phoxinus phoxinus* (Cyprinidae, Ostariophysi, Pisces). J. Comp. Physiol. 82: 407–410.
Pfeiffer, W. & G. Riegelbauer. 1978. The effect of the alarm substance on the central nervous excitation of the black tetra *Gymnocorymbus ternetzi* (Characidae, Ostariophysi, Pisces) indicated by dorsal light response. J. Comp. Physiol. 123: 281–288.
Salyer, J.C. & K.F. Lagler. 1940. The food habits of the American merganser during winter in Michigan, considered in relation to fish management. J. Wildl. Manage. 4: 189–219.
Salyer, J.C. & K.F. Lagler. 1946. The eastern belted kingfisher *Megaceryle alcyon alcyon* (Linnaeus), in relation to fish management. Trans. Amer. Fish. Soc. 76: 97–117.
Seghers, B.H. 1974. Schooling behavior in the guppy (*Poecilia reticulata*): an evolutionary response to predation. Evol. 28: 486–489.
Shaw, E. 1978. Schooling fishes. Amer. Sci 66: 166–175.
Smith, R.J.F. 1977. Chemical communication as adaptation: alarm substance of fish. pp 303–320. *In*: D. Müller-Schwarze & M.M. Mozell (ed.) Chemical Signals in Vertebrates, Plenum Press, New York.
Thines, G. & E. Vandenbussche. 1966. The effects of alarm substance on the schooling behavior of *Rasbora heteromorpha* Duncker in day and night conditions. Anim. Behav. 14: 296–302.
Van Olst, J.C. & J.R. Hunter. 1970. Some aspects of the organization of fish schools. J. Fish. Res. Board Can. 27: 1225–1238.
Verheijen, F.J. 1956. Transmission of a flight reaction amongst a school of fish and the underlying sensory mechanisms. Experientia 12: 202–204.
Verheijen, F.J. 1962a. Gas spitting by alarmed fish disturbs their hydrostatic equilibrium. Science 137: 864–865.
Verheijen, F.J. 1962b. Alarm substance and intraspecific predation in cyprinids. Die Naturwissenschaften 49: 356.
Verheijen, F.J. 1963. Alarm substances in some North American cyprinid fish. Copeia 1963: 174–175.
Verheijen, F.J. & J.H. Reuter. 1969. The effect of alarm substance on predation among cyprinids. Anim. Behav. 17: 551–554.
Williams, G.C. 1964. Measurement of consociation among fishes and comments on the evolution of schooling. Publ. Mus. Mich. State Univ. Biol. Series 2: 349–384.

Env. Biol. Fish. Vol. 6, No. 1, pp. 31–38, 1981

Attraction of zebrafish, *Brachydanio rerio*, to isolated and partially purified chromatographic fractions

Faye D. Algranati[1] & Alfred Perlmutter

Laboratory of Aquatic Biology, Department of Biology, Graduate School of Arts & Science, New York University, New York, N.Y. 10013, U.S.A.

Keywords:
Pheromone, Intrasexual, T-maze, Thin-layer chromatography, Roto-evaporation, Lipid

Synopsis

Water from donor fish of either sex maintained in tank systems for 16 days was tested to determine intrasexual responses in a T-maze apparatus. Only donor water attracted fish, suggesting the presence of intrasexual pheromone(s). The sexual attractant(s) was removed by methylchloroform extraction. The residue from this extraction elicited positive responses in test fish. Dried residues showed 5 bands in thin-layer chromatograms, but only one band (R_f 0.94), identified as cholesterol ester, contained the sexual attractant(s).

Introduction

Pheromones are exogenous secretions which influence many aspects of physiology and behavior. While the concept of chemical communication as a means of sexual discrimination was suggested in the past (Jaski 1939), accumulating evidence (Davis et al. 1974, Bloom & Perlmutter 1977) has shown that pheromones can be a factor in sexual attraction in fish. Although chemical characterization of sex pheromones has been achieved in insects (Meinwald et al. 1971, Tumlinson et al. 1971), and some mammals (Michael et al. 1971, Epple 1976) such information is not available for any fish.

Sex pheromones have been reported in many fish, including the goby *Typhlogobius californiensis* (MacGintie 1939), *Bathygobius* (Tavolga 1956), channel catfish *Ictalurus punctatus* (Timms & Kleerekoper 1972), paradise fish *Macropodus opercularis* (Davis et al. 1974), and goldfish *Carassius auratus* (Partridge et al. 1976). Chemical characterization has been achieved in none of these species.

Recently, Bloom & Perlmutter (1977, 1978) have demonstrated both intra- and intersexual sex pheromones in zebrafish, *Brachydanio rerio*. Attraction was dependent upon the time a number of fish were kept in a tank. Characterization of the attractant(s) was not attempted, but it (they) could be removed from the water by charcoal filtration. This study was a continuation of the investigation of the intrasexual attractant(s), primarily in the isolation and partial purification.

Materials and methods

Prior to experimental procedures, 120 mature male and female zebrafish, *Brachydanio rerio* (Hamilton-Buchanan) (Amazon Aquarium, Brooklyn, New York), were kept separately in 38 liter holding tanks

[1] Current Address: 264 East Broadway, New York, N.Y. 10002, U.S.A.

Received 28.12.1979 Accepted 29.8.1980

© Dr. W. Junk b.v. Publishers, The Hague. Printed in The Netherlands. 0378-1909/81/061-031 $01.60.

at concentration of 10 fish per 33 liters of water. Tank water was aerated and filtered through glass wool (Corning Glass works, Corning, New York). Experimental fish were kept for 16 days in 38 liter capacity aquaria at concentrations of 20 fish per 33 liters of water. The fish in this water which was used for assay for the presence of sexual attractant(s) were designated 'donor' fish and the water as 'donor' water. This donor water was tested in the T-maze for the presence of sexual attractant(s). All fish were kept in a temperature controlled room (23-24° C) and exposed to 15 h daylight photoperiod (0630-2130 h).

Aquarium systems

The aquarium systems were four 38 liter tanks each containing 33 liters of water and connected to each other by 'U' shaped glass siphons. Water from the first tank, where 20 donor fish of either sex were kept, was returned to the fourth tank by an air lift. Flow rate into the fourth tank was adjusted to 300-400 ml· min^{-1}. The water was then returned, by siphoning action, via the third and second tanks to the first tank.

All fish were maintained in New York City tap water modified by passage through a deionizer cartridge (D0809, Barnstead Spring Co., Boston, Mass.). For every 171 liters of deionized water, 19 liters of untreated tap water and 5 mm of sodium bicarbonate were added to maintain the pH at 7.0. In addition, 0.33 g of sodium chloride and 0.05 g of calcium chloride were added to each liter of prepared water.

Diet

Since it has been reported in insects that sex attractant pheromones are partially or completely dependent on food source (Borden et al. 1969, Birch 1978), a series of experiments were conducted to determine whether the sexual attractant(s) in the zebrafish could also be affected by diet.

All fish were maintained on one of two diets for 1 month prior to experimentation. One group of males and females received a high lipid diet consisting of 0.2 g ground Purina Dog Chow (at least 8% fat), and live *Daphnia*. A second group of fish received a low lipid diet, (0.2-0.6% fat), which consisted of 0.2 g of ground fillet of blackback flounder, *Pseudopleuronectes americanus* (Tressler & Lemon 1951), and live *Daphnia*. Both groups were fed daily throughout the week.

At the end of this period, 20 donor fish which had been fed the high lipid diet, were placed in a four 38 liter tank system and continued to be fed the same diet. Another 20 fish which had been fed the low lipid diet were placed in a similar system and continued to be fed the same diet. The fish were maintained in the systems for 16 days. The control consisted of a similar tank system without fish into which the same amount of the appropriate food was introduced daily with the exception of *Daphnia* for a 16 day period. After this time interval, fish were tested in the T-maze for the possible interaction of diet with the activity of intrasexual attraction.

Extraction

Removal of the sexual attractant(s) from donor water of the SU of either sex has been accomplished with activated charcoal (Bloom & Perlmutter 1977). Difficulty in elution of possible sexual attractant(s) from the charcoal precluded further study. Since methylchloroform has been described as effective in the extraction of growth-inhibiting substances in fishes (Yu & Perlmutter 1970, Pfunderer et al. 1974, Heller 1976) this organic solvent was used in an attempt to extract the sexual attractant(s) in the zebrafish.

The extraction unit was a 2000 ml Erlenmeyer flask containing 500 ml of methylchloroform overlayed with 1100 ml of deionized water. A glass-coated stirring bar insured continuous mixing. A long-stemmed funnel was placed in the flask, with the cone resting on the lip of the flask, and the distal end submerged in the methylchloroform. A piece of glass wool was inserted in the cone to collect debris. The unit was then placed on a magnetic stirrer and rotation of the stirring bar adjusted to 40 revolutions per minute.

Water from the tank containing the donor fish was pumped into the funnel, through the stem of the funnel into the methylchloroform where it then rose above the solvent layer, and passed through siphons into the fourth tank of the system. Water was returned to the first tank via siphon action.

Initially, 20 fish of either sex were maintained in a four 38 liter tank system and the water constantly circulated for 16 days through methylchloroform contained in the extraction unit. It was found that droplets of methylchloroform were carried from the extraction unit into the first tank and resulted in a noticeable weight loss in the donor fish.

To avoid such detrimental effects, water was circulated in a tank system without extraction. After 16 days, the fish were removed and the water circulated through an extraction unit containing methylchloroform for 72 h.

To test for the presence of intrasexual attractant(s) in donor water subjected to methylchloroform extraction, donor water from either sex, after the prescribed hours of extraction, was tested in the T-maze with fish of the same sex as the donors. In another series of experiments, tests were made for the presence of intrasexual attractant(s) in the methylchloroform extractant. The residue after roto-evaporation of the extractant was redissolved in water equivalent of a four 38 liter tank system and the presence of the intrasexual attractant(s) assayed in the T-maze.

Thin-layer chromatography (TLC)

At the end of the extraction period the extracted material was prepared for chromatographic development by roto-evaporation of the methylchloroform, which left behind a yellow-brown residue. This residue was redissolved in 2 ml of methylchloroform, dried under nitrogen, weighed and stored at —20° C until needed. Prior to application of the residue to the chromatographic plates, a calculated amount of methylchloroform was added for a concentration of 300 micrograms per 5 microliters.

Commercially prepared TLC plates (20 × 20 cm), precoated with a 0.25 mm thick layer of Silica Gel G, were used (AnalTech Inc., Newark, I11) in an attempt to isolate and purify the sexual attractant(s). Samples dissolved in methylchloroform were applied in five microliter volumes, 2 cm from one edge of the plate. A lipid standard (Fisher Scientific Co., Springfield, N.J.) consisted of monoglyceride, 1,2 and 1,3 diglycerides, fatty acid, triglyceride and cholesterol ester, were applied on all TLC plates. This was used as a basis of comparison with the substances isolated from the methylchloroform solubilized material originating in donor waters. Several plates were utilized to factionate the entire methylchloroform-redissolved sample.

Prior to the development of the TLC plates containing the redissolved residues and lipid standard, a glass chromatograph tank (30 × 10 × 26 cm) was allowed to equilibrate for one hour with a solvent mixture of ethyl acetate: glacial acetic acid: acetone:heptane (6.5:1.5:2.5:39.5 volume per volume) (Kates 1972). The plates were placed in the chromatograph tank and the solvent front allowed to ascend 10 cm beyond the origin. The chromatogram was removed, airdried for 10 min, and sprayed with one of two developing reagents.

The first reagent consisted of 20 g ammonium sulphate and 48 ml of sulphuric acid per 100 ml of water. The reagent was sprayed on the plates which were then placed in a pre-heated 200° C oven for 1h. At the end of this time, bands were visualized and R_f values calculated.

The second reagent employed to test for the presence of cholesterol and its esters consisted of concentrated sulfuric acidacetic acid solution (1:1 volume per volume) and was applied to dried TLC plates. Treated plates were placed in a preheated 90° C oven for 15 min at which time cholesterol and its esters, if present, would appear as red spots (Jatzkewitz & Mehl 1960).

Finally, in order to test for the presence of sexual attractant(s) in the observed bands, the portion of the plate which contained the lipid standard was charred with the first reagent and used as a guide for the uncharred portion of the plate. The uncharred gel was then removed, centrifuged in 1 ml of a chloroform-methanol solution 3 times, and the washes pooled. Ten milliliters of the pooled wash was added to each carboy containing 18 liters of control water. Each band was treated separately, and the redissolved individual bands from both sexes were then tested in the T-maze for intrasexual attraction.

T-maze testing system

The T-maze used in the experiments was a smaller

version of the one used by Bloom & Perlmutter (1977). Following their testing technique experiments were conducted under low diffuse light to eliminate the casting of shadows into the T-maze (Breder & Halpern 1946). Individual fish were kept within the pen area and prevented from leaving by a plastic wire mesh screen located inside the T-maze. Experimental and control waters entered the T-maze from 20 l glass carboys that fed by siphon action into regulating, constant flow boxes. Water was calibrated to flow from the constant flow boxes to each arm of the T-maze at a rate of 650 $\pm$ 3 ml$\cdot$min^{-1}. After a 5 min acclimation period, during which time the fish was subjected to a constant flow of experimental and control water from both arms of the T-maze, the screen was lifted. Insertion of the fish's head into an arm of the T-maze was considered a positive reaction. Each experiment included three groups of eight fish (24 fish) of each sex and required approximately 90 liters of experimental water and an equivalent amount of control water.

Each complete series of 24 fish was statistically analyzed using chi-square adjusted for bias by the Yates Correction for Continuity. The null hypothesis was that there was no difference between the experimental or control waters; this was characterized as randomness of choice between the T-maze arms. However, the null hypothesis was rejected if the experimental fish entered one arm of the maze more frequently (two-way test).

Results

The intraspecific reaction of males and females to non-extracted donor water from their respective sex is given in Table 1. Both sexes showed a significant attraction to donor water ($p<0.01$). Nonextracted donor water, containing no fish but with aliquots of either food, did not attract wither sex. However, nonextracted donor water from males and females fed these diets attracted fish of the same sex (Table 2). No preference was shown by either males or females in the choice between methylchloroform-extracted donor water of their own sex and control water (Table 3).

After a 72 h methylchloroform extraction of donor water, the calculated amount of residue from males was 53.3 $\pm$ 1.6 μg, and from females 55.2 $\pm$ 2.5 μg. The redissolved extracted residues elicited intrasexual responses ($p<0.01$) for both sexes (Table 4). Thin-layer chromatographic plates developed in prescribed solvent mixtures showed 5 major bands in both dog food and zebrafish sexual attractant(s) residues, and 6 bands in the lipid standard (Fig. 1). The major charred bands (1, 2, 3, 4 and 6) from the collected residues and dog food had similar migration patterns to the lipid standard. The calculated R_f values for the 5 bands were 0.08, 0.34, 0.53, 0.75 and 0.94, consecutively.

The presence of cholesterol ester was indicated from residue samples of both dog food and donor waters and in the lipid standard when the plates were treated with a sulphuric acid-acetic acid solution. The red coloration indicated the presence of this substance only in band 1 (R_f 0.94) which corresponded to the cholesterol ester fraction of the lipid standard. Silica gel washes and elutants of each of the 5 bands obtained from the food components did not elicit intrasexual attraction. This was also true for bands 2-6 obtained from the residue of extracted donor water. However, band 1 proved to be the single one which did elicit intrasexual attraction ($p<0.01$) (Tables 5, 6).

Table 1. Male reaction to nonextracted male donor water, and female reaction to nonextracted female donor water.

Donor water	Sex of reactant fish	Tank system	No. of fish	No. of days	To donor water	To control water	Total to donor water	Total to control water	Chi-square value	P value
Male	Male	4	20	16	6	2				
					6	2	18	6	5.042	>0.01
					6	2				
Female	Female	4	20	16	6	2				
					6	2	18	6	5.042	>0.01
					6	2				

Table 2. Male and female reaction to nonextracted water receiving daily aliquots of food of low or high lipid content, compared to their reaction to nonextracted water from fish, of their respective sex, fed these foods.

	No. of fish	No. of days	Tank system	To lipid water	To donor water	Total to lipid water	Total to donor water	Chi-square value	P value
Male:									
Low	20	16	4	2	6				
lipid				2	6	6	18	5.042	>0.01
food*				2	6				
High									
lipid									
food*	20	16	4	2	6				
				2	6	6	18	5.042	>0.01
				2	6				
Female:									
Low	20	16	4	2	6				
lipid				1	7	5	19	7.042	<0.01
food				2	6				
High									
lipid	20	16	4	2	6				
food				1	7	5	19	7.042	<0.01
				2	6				

* Daily aliquots consisted of 0.2 g of flounder and *Daphnia.*
** Daily aliquots consisted of 0.2 g of ground Purina Dog Chow and *Daphnia.*

Table 3. Male reaction to male extracted donor water and female reaction to female extracted donor water in the four tank system.

Donor water	Sex of reactant fish	Tank system	No. of fish	No of days	To donor water	To control water	Total to donor water	Total to control water	Chi-square value	P value
Male	Male	4	20	16	4	4				
					4	4	12	12	0.00	N.S.
					4	4				
Female	Female	4	20	16	4	4				
					4	4	12	12	0.00	N.S.
					4	4				

N.S. = no significant.

Table 4. Male and female reaction to redissolved donor residue of the same sex.

Sex of fish	To residue water	To control water	Total to residue water	Total to control water	Chi-square value	P value
Male	6	2				
	6	2	18	6	5.042	>0.01
	6	2				
Female	6	2				
	6	2	18	6	5.042	>0.01
	6	2				

Discussion

Bloom & Perlmutter (1977) demonstrated both intra- and intersexual attraction in zebrafish. They suggested the active pheromone(s) may function in schooling or spawning behavior. In the present study, nonextracted donor water of males and females demonstrated intrasexual attraction, suggesting the presence of an intrasexual pheromone(s)

The possible interaction of diet with the activity of the intrasexual pheromone(s) was not fully investigated by Bloom & Perlmutter (1977). Present results indicate that with at least these two different lipid diets, attractiveness of donor water was not altered, nor did the food regime itself act as a sexual attraction factor. These findings reinforce the concept that the release of exogenous factors by the zebrafish is responsible for the attraction.

Methylchloroform extraction proved effective in removing the pheromone(s) from donor water of both sexes, considering test fish did not respond to extracted donor water. As methylchloroform removed the attractant(s), it was anticipated that the active material(s) would be present in the residue following rotoevaporation, and might remain biologically active. This proved to be the case. Isolation and partial purification by thin-layer chromatography showed the active substance(s) was located in band 1. But band 1 from an extract of food added to a tank system, did not attract test fish. These results suggest that band 1, the cholesterol ester band, detected in food and the sexual attractant(s), migrates to the same position on the TLC plate. But the food and sexual attractant(s) cholesterol esters are two different substances that only happen to migrate to the same position on the plate under experimental conditions. This is the first report in which an active band has been isolated and partially identified as a sexual pheromone in any fish.

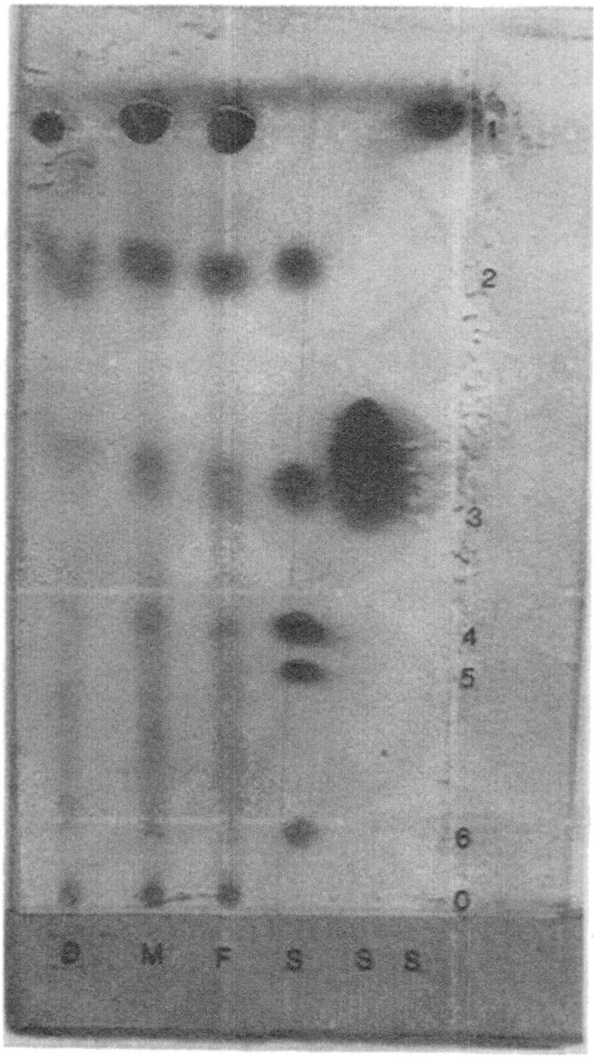

Fig. 1. Thin-layer chromatographic plate. D = dog food (high lipid content), M = male donor residue, F = female donor residue, S = Standards. Band 1 = cholesterol ester (R_f 0.94), Band 2 = triglyceride (R_f 0.73), Band 3 = fatty acid (R_f 0.53), Band 4 = 1,3 diglyceride (R_f 0.34), Band 5 = 1,2 diglyceride (R_f 0.28), Band 6 = monoglyceride (R_f 0.08), and O = origin.

Table 5. Male reaction to fraction of processed male donor water[a] by silica gel washes of individual bands developed on TLC plates.

No. of days	No. of fish	Band[b] No.	To donor wash water fraction	To control water	Total to donor wash water fraction	Total to control water	Chi-square value	P value
16	20	1	6	2				
			6	2	18	6	5.042	>0.01
			6	2				
16	20	2	4	4				
			3	5	11	13	0.042	N.S.
			4	4				
16	20	3	4	4				
			4	4	12	12	0.00	N.S.
			4	4				
16	20	4	5	3				
			3	5	13	11	0.042	N.S.
			5	3				
16	20	6	4	4				
			4	4	12	12	0.00	N.S.
			4	4				
16	20	0	5	3				
			5	5	13	11	0.042	N.S.
			3	5				

Table 6. Female reaction to fraction of processed female donor water[a] by silica gel washes of individual bands developed on TLC plates.

No. of days	No. of fish	Band[b] No.	To donor wash water fraction	To control water	Total to donor wash fraction	Total to control water	Chi-square value	P value
16	20	1	6	2				
			6	2	18	6	5.042	>0.01
			6	2				
16	20	2	5	3				
			4	4	13	11	0.042	N.S.
			4	4				
16	20	3	4	4				
			4	4	12	12	0.00	N.S.
			4	4				
16	20	4	4	4				
			4	4	12	12	0.00	N.S.
			4	4				
16	20	6	4	4				
			4	4	12	12	0.00	N.S.
			4	4				
16	20	0	4	4				
			4	4	12	12	0.00	N.S.
			4	4				

[a] Donor water was obtained from tanks which held 20 male fish for 16 days. After extraction with methylchloroform the solvent was rotoevaporated and the residue redissolved in methylchloroform. Five microliter samples of a methylchloroform donor extract was applied at the origin of each of several TLC plates. After migration of each sample, in the prescribed solvent mixture, the individual bands were eluted from the silica gel with a chloroform-methanol solution. The washes were pooled and 10 ml of the chloroform-methanol solution added to each 18 liters of water of several carboys.
[b] Fraction of donor wash water visible as a band on TLC plates.

References cited

Birch, M.C. 1978. Chemical communication in pine bark beetles. Amer. Sci. 66: 409–419.

Bloom, H.D. & A. Perlmutter. 1977. A sexual aggregating pheromone system in the zebrafish, *Brachydanio rerio* (Hamilton-Buchanan). J. Exp. Zool. 199: 215–226.

Bloom, H.D. & A. Perlmutter. 1978. Possible pheromone mediated reproductive isolation in two species of cyprinid fishes of the genus *Brachydanio*. J. Fish Biol. 13: 47–50.

Borden, J.H., KK. Nair & C.E. Slater. 1969. Synthetic juvenile hormone: induction of sex pheromone production in *Ips confusus*. Science 166: 1626–1627.

Breder Jr., C.M. & F. Halpern. 1946. Innate and acquired behavior affecting the aggregation of fishes. Physiol. Zool. 19: 154–192.

Davis, R.E., C. Harris. & J. Shelby. 1974. Sex differences in aggressivity and the effects of social isolation in the anabantoid fish, *Macropodus opercularis*. Behav. Biol. 11: 497–509.

Epple, G. 1976. Chemical communication and reproductive processes in non-human primates pp. 257–282. *In*: Mammalian Olfaction, Reproductive Processes and Behavior. American-Elsevier, New York.

Heller, R.F. 1976. The role of abiotic and biotic factors in the environment on growth in the zebrafish, *Brachydanio rerio* (Hamilton-Buchanan). Ph.D. Thesis, New York University, New York, 100 pp.

Jaski, C. 1939. Ein Oestruszyklus bei *Lebistes reticulatus* (Peters). Proc. Koninkl. Nederl. Akad. Wetensch. 42: 201–207.

Jatzkewitz, H. & E. Mehl. 1960. Thin-layer chromatography: identification of cholesterol and its derivatives. Z. Physiol. Chem. 320: 251–260.

Kates, M. 1972. Techniques of lipidology. American-Elsevier, New York, 610 pp.

MacGinitie, G.E. 1939. The natural history of the blind goby, *Typhlogobius californiensis* (Steindachner). Amer. Midl. Nat. 21: 489–500.

Michael, R.P., E.P. Keverne & R.W. Bonswall. 1971. Pheromones: isolation of male sex attractants from the female primate. Science 172: 964–966.

Partridge, B.L., N.R. Liley & N.E. Stacey. 1976. The role of pheromones in the sexual behavior of the goldfish. Anim. Behav. 24: 291–299.

Pfunderer, P., P. Williams & A.A. Francis. 1974. Partial purification of the crowding factor in *Carassius auratus* and *Cyprinus carpio*. J. Exp. Zool. 187: 375–382.

Tavolga, W.N. 1956. Visual, chemical and sound stimuli as cues in sex descriminatory behavior of the gobiid fish *Bathygobius soporator*. Zoologica 41: 49–64.

Tressler, D.K. & J. McW. Lemon. 1951. Marine products of commerce: their acquisition, handling, biological aspects and the science and technology of their preparation and preservation. Reinhold Pub. Corp., New York. 728 pp.

Tumlinson, J.H., R.C. Gueldner, D.D. Hardee, A.C. Thompson, P.A. Heden & J.F. Minyard. 1971. Identification and synthesis of the four compounds comprising the boll weevel sex attractant. J. Org. Chem. 36: 2616–2620.

Yu, M.L. & A. Perlmutter. 1970. Growth-inhibiting factors in the zebrafish, *Brachydanio rerio*, and the blue gourami, *Trichogaster trichopterus*. Growth 35: 153–175.

Env. Biol. Fish. Vol. 6, No. 1, pp. 39–48, 1981

Reduction of competition between bisexual and unisexual females of *Poecilia* in northeastern Mexico

Joseph S. Balsano[1], Kristine Kucharski[1], Edward J. Randle[1], Ellen M. Rasch[2] & Paul J. Monaco[2]

[1]*Division of Science, University of Wisconsin-Parkside, Kenosha, Wisconsin 53141, U.S.A.*
[2]*College of Medicine, East Tennessee State University, Johnson City, Tennessee 37614, U.S.A.*

Keywords:
All-female, Gynogenesis, Habitat preference, Mate preference, Niche dimensions, Resource allocation, Social interaction, Sympatric coexistence, Superfetation, Triploidy

Synopsis

Breeding compexes of poeciliid fishes with a bisexual and two unisexual species were studied for mechanisms permitting sympatric coexistence. The unisexuals are gynogenetic and thereby sexually dependent on the males of the bisexual species for sperm to initiate development, but inheritance is entirely maternal.

Bisexual females are more abundant in headwater localities; unisexuals increase in downstream localities. Males were 10 – 18% of the total poeciliid population, regardless of the relative proportions of bisexual to unisexual females. Downstream localities were typified by greater habitat diversity, including a variety of backwater pools. The unisexuals showed a marked preference for such pools. Both field and laboratory studies showed that all three types of females as well as males preferred shaded areas with a gravel substrate.

Although the four types of fish were found together, nearest neighbour data indicated that each type of female preferred its own kind. Males courted throughout the year and were indiscriminate in their choice of mates. Despite the skewed sex ratio, males were not in short supply because only a few females were sexually receptive at a given time. No significant differences existed between bisexuals and unisexuals in their relative reproductive outputs, but they were asynchronous.

Introduction

Within the arroyos and intermittent streams of northeastern Mexico exist breeding complexes of both bisexual and unisexual species of some live-bearing fishes (commonly called mollies). The two unisexual female species are gynogenetic and therefore dependent upon the males of the bisexual species for reproduction. The unisexual females produce eggs with a somatic chromosome complement (Rasch & Balsano 1974). These eggs require sperm from a congener to initiate development, but inheritance is entirely maternal. Although the three types of females compete for a common sperm source, they coexist in great numbers throughout their range (Balsano et al. 1972, Rasch & Balsano 1974). All three types are closely related (Fig. 1). One unisexual, *Poecilia formosa*, is a diploid hybrid of two bisexual species, *P. latipinna* and *P. mexicana*, that it sexually parasitizes. The other unisexual species is a triploid hybrid that occurs when sperm from males actually fertilize the diploid eggs of *P. formosa*.

Received 15.2.1980 Accepted 29.8.1980

© Dr. W. Junk b.v. Publishers, The Hague. Printed in The Netherlands. 0378-1909/81/061-039 $02.00.

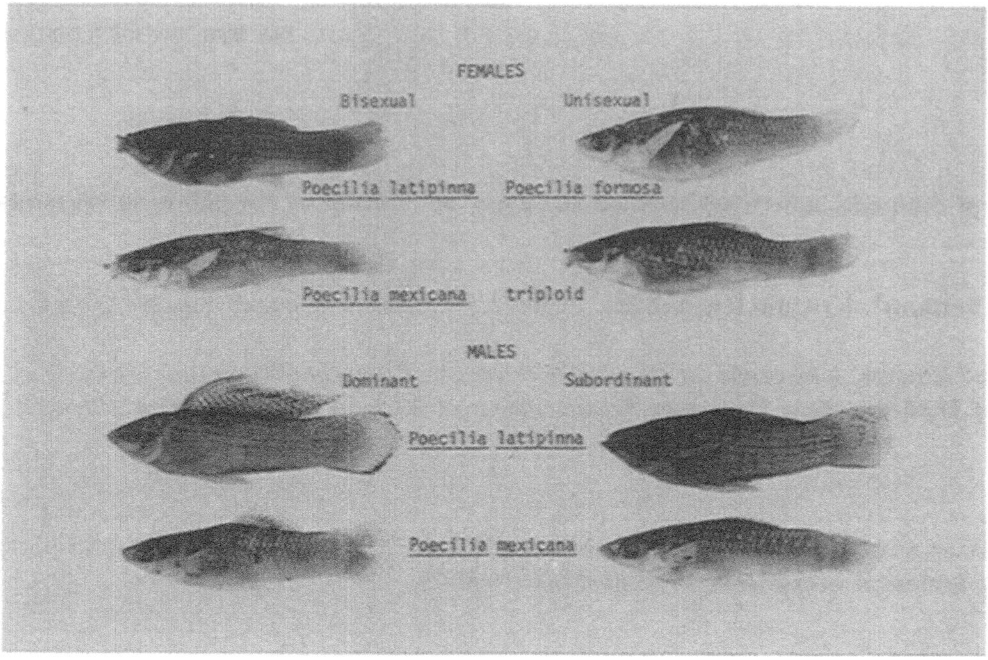

Fig. 1. Members of bisexual and unisexual complexes of *Poecilia* found in NE Mexico and SE Texas. This report is concerned only with those members that are found in the tributaries of the Soto la Marina drainage of NE Mexico, i.e., the females and males of bisexual species, *P. mexicana*, and the unisexual females of *P. formosa* and its triploid associate.

This complex offers an opportunity for the study of competitive exclusion (Hardin 1960, Hutchinson 1965), because of the dependent relationships among the unisexual and bisexual species. Their close genetic relationship implies similar if not identical ecological requirements. The purpose of this paper is to describe several niche dimensions in which competition has been reduced.

Materials and methods

Relative frequencies for the species mix were based upon more than 10,000 specimens collected with 0.6 cm seine during 12 field trips to 19 locations in the Soto la Marina drainage of northeastern Mexico (Figs. 2, 3). Since most of our field data came from the Rio Purificacion, this river was selected as the best tributary to test our hypothesis that the frequencies of unisexuals increased in downstream localities. Two to six more samples were collected during 1977 and 1978 from five localities in the Rio Purificacion and combined with those collected from 1970 to 1975 (Fig. 2). In addition, samples were collected from four new sites during 1977 and 1978: two samples from Santa Cruz, three each from Las Cruces and site X, and one sample from site Z. Data from all these Rio Purificacion collections were combined and are summarized in Figure 3.

Habitat preference and social interactions were studied by using fish identified by color-coded tags. After each fish was identified on the basis of morphological and cytological criteria, a tiny incision was made in the epaxial musculature just anterior to the insertion of the dorsal fin for insertion of the tag (Kucharski 1979). The 8 mm long tags were cut from three colors of Dymo Labeling Tape: yellow for the bisexual females of *P. mexicana*, orange for triploid unisexuals, and pale green for diploid unisexuals of *P. formosa*. Fish were healthy and robust even after three months with tags.

Habitat preference experiments were carried out sequentially in the laboratory and in the field. The initial experiments were conducted in four rectangular pools, 1.2 m wide by 2.4 m long, located in a greenhouse. Half of each pool had a sand substratum and the other half had large gravel; in addition, half of each substratum was shaded. Thus, four zones of equal size were established with

5 to 10 cm depth of water and airstones were used to circulate water to avoid temperature gradients. Ten of each type of female and three males were placed in each pool and their distributions were recorded daily at 0900, 1200 and 1500 h for three months. A second series of laboratory experiments were conducted using only sand in pool A, only gravel in B, both types in C and D, and only 5 females of each type per pool and no males. Data were recorded only at 1200 h since field observations indicated

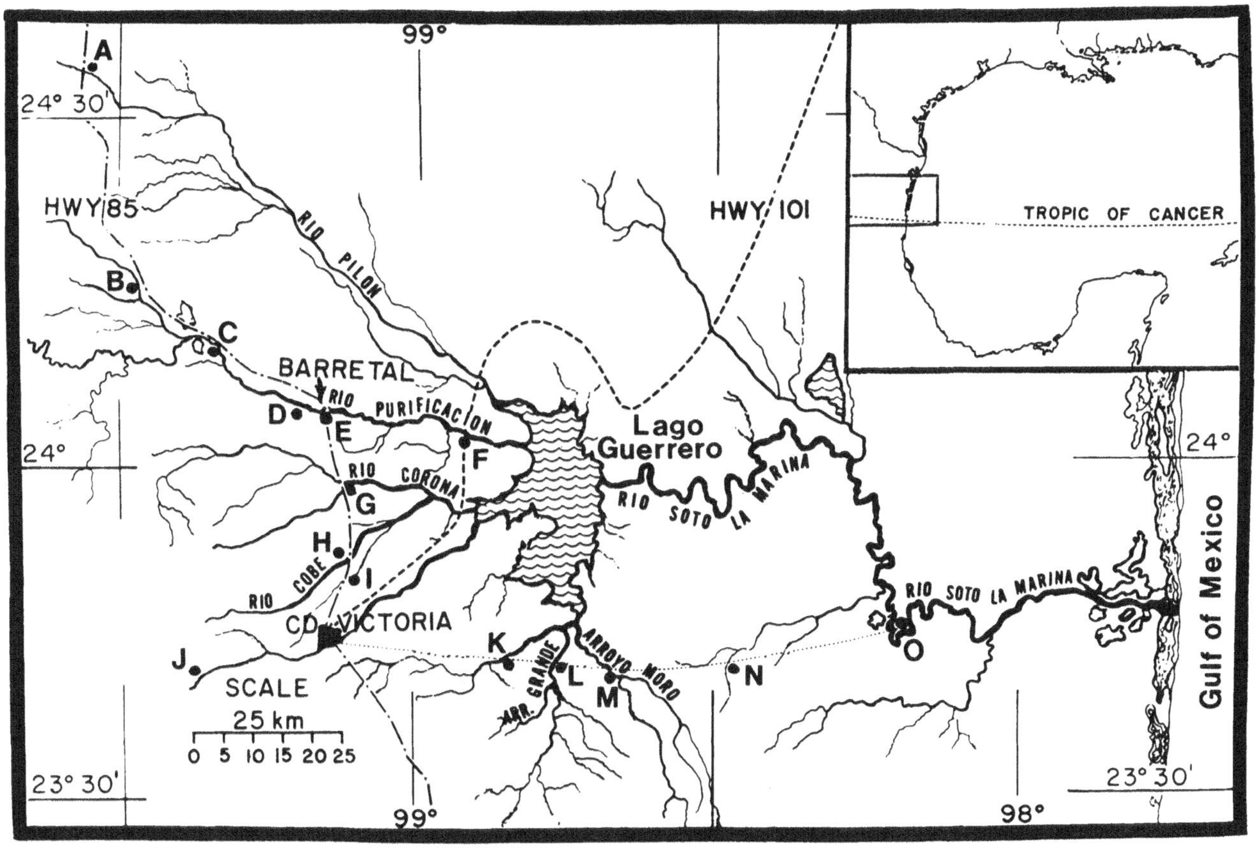

Fig. 2. Percentages of unisexuals among females of *Poecilia* spp. found in fifteen locations in the Soto la Marina drainage of northeastern Mexico. Data are based upon multiple samples collected during one to nine field trips including spring, summer, and winter from 1970 through 1975 to each location as follows:

Letter and location	Number of collections	Total No. females	Percentage of unisexuals *P. formosa*	triploids	totals
A: Rio Pilon	3	101	0.0	1.0 ± 1.0	1.0
B: Rio Purificacion at Hidalgo	8	385	0.8 ± 0.4	3.9 ± 1.0	4.7
C: Rio Purificacion at Palo Alto	3	176	1.1 ± 0.8	2.3 ± 1.1	3.4
D: Rio Purificacion at earthen dam	1	39	7.7 ± 4.3	7.7 ± 4.3	15.4
E: Rio Purificacion at Barretal	9	2125	12.3 ± 0.8	38.2 ± 1.1	50.5
F: Rio Purificacion at N. Padilla	9	1476	32.2 ± 1.2	20.3 ± 1.0	52.4
G: Rio Corona	9	518	0.6 ± 0.3	19.3 ± 1.7	19.9
H: Rio Cobe	9	541	0.4 ± 0.3	6.8 ± 1.1	7.2
I: Arroyo la Presa	9	915	1.7 ± 0.4	13.2 ± 1.1	15.0
J: Rio San Marcos	1	8	0.0	0.0	0.0
K: Arroyo Sarnoso	2	54	0.0	14.8 ± 4.8	14.8
L: Arroyo Grande at Casas	8	296	60.1 ± 2.8	33.8 ± 2.7	93.9
M: Arroyo Moro	9	2139	2.2 ± 0.3	14.9 ± 0.8	17.1
N: Rio Piriuli	1	23	0.0	17.4 ± 7.9	17.4
O: Rio Soto la Marina	1	4	0.0	100.0	100.0
Grand totals for all locations:		8800	11.3 ± 0.1	20.7 ± 0.4	32.0

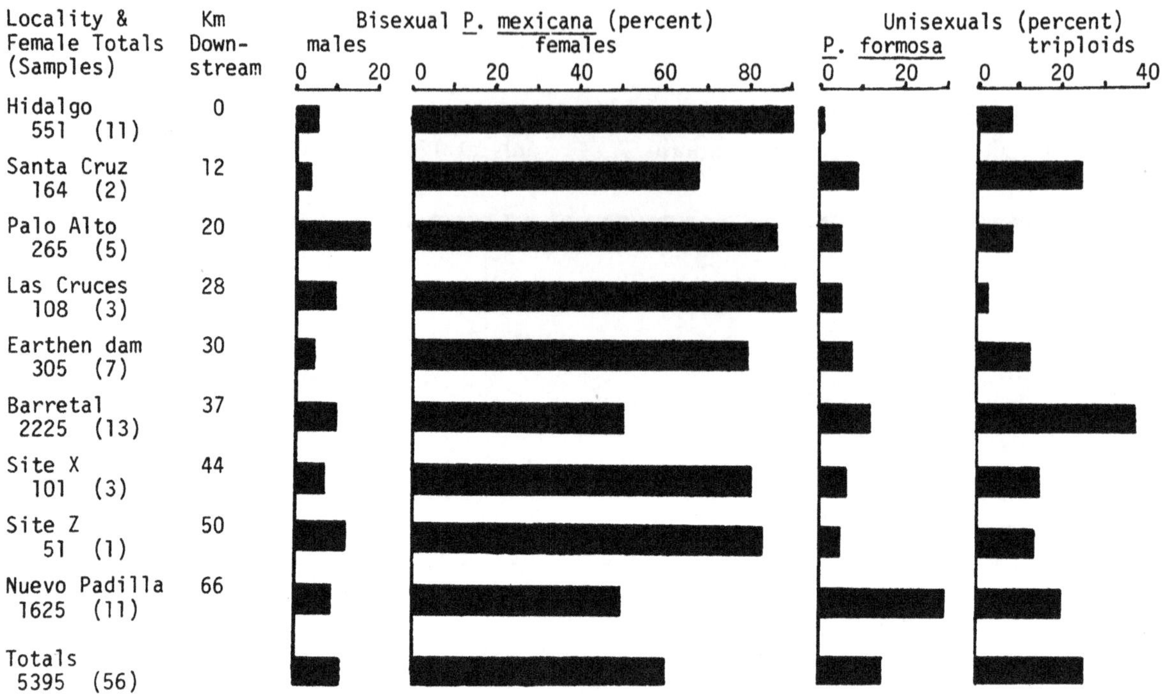

Fig. 3. Relative frequencies of bisexual and unisexual *Poecilia* collected from nine different localities along a 66 km segment of the Rio Purificacion which is a major tributary of the Rio Soto la Marina of NE Mexico. Each locality was sampled from one to 13 times during 1970 to 1978; this information is shown in parentheses following the total number of females that were collected from each locality. The percentages for each type of female are based on these sample totals; the percentages of males, however, are based upon the combined totals of males and females. Consequently, the percentages for all the females equal 100%.

that the fish were most active at this time. The field experiment was conducted in a shallow, isolated pool about 6 m by 4 m at the earthen dam on the Rio Purificacion (Fig. 2). This pool was divided into eight zones: six had large gravel and rocks whereas only two had sand; a tarpaulin supported by tent poles shaded two of these zones. Twenty-four females were placed in this pool and observed for a total of 19 hours in 16 separate observations.

Female preference experiments were conducted in a 15 m by 4 m isolated pool at site X on the Rio Purificacion (Fig. 2, 3) using 20 females of each type and no males. All resident adult fish were removed from this pool prior to the introduction of the 60 tagged fish. Two kinds of data were recorded: 39 different observations were made over a five day period and tallies were kept of which females were the nearest neighbours of a given female; over a two hour period on one day 20 groups of fish were photographed on 35 mm camera film. From the slides fish were enlarged to 7 cm on a screen. The distances between the points of insertion of the fish tags were measured for all females within two standard lengths of a reference fish. This procedure was repeated for all the fish on each slide. The distances for the six different paired combinations were than statistically analyzed.

Mate preference experiments were conducted with sexually mature fish collected from the field. These fish were kept over a year in community aquaria (containing both males and females) except when they were isolated for a specific experiment. A male and two equally-receptive females were used in these experiments. Female receptivity was determined on the basis of size, aggressiveness, current gravida status, and at least one week since most recent delivery of a brood. An 85 l aquarium was divided into three compartments for mate preference experiments. The male was isolated in the middle compartment with a female in each of the end compartments. The partitions between compartments were made of translucent plexiglass with many tiny holes to permit water flow among compartments. The three isolated fish were allowed to acclimate over the weekend and then 8 or 10 observations of 5 min each were made over the next five days. A string and pulley mechanism was used to raise the two partitions to avoid frightening the fish. Two kinds of data were recorded: how many seconds were spent directly courting a female and how many times the male thrusted his gonopodium into the female's genital pore during a five minute observation period. Some mate preference observations were also made in conjunction with other field experiments.

Comparisons of reproductive output of the different types of females were based upon fish collected in April 1971 and 1972 and in June 1970 and July 1972. Samples that included both bisexual and unisexual females were obtained from all localities shown in Figure 2 except sites C, J, and O. The proportion of unisexuals in these samples varied from 2% to 94% of the females in a given sample. The numbers of females examined, and proportions of unisexuals are shown in Table 4 together with the results of the calculations of reproductive output. The reproductive output per kind of female was calculated as the product of percentage of females with embryos times the mean number of embryos per female.

Results

The frequencies of the three types of females exhibited considerable spatial and temporal fluctuations, but generally the bisexual females were more abundant in headwater localities whereas unisexuals increased downstream (Fig. 2). Sites A, B, C and J (headwater habitats) all had less than 5% unisexuals, while F and L (downstream) had the highest percentages of unisexuals. Ecologically intermediate sites typically had 15 – 20% unisexuals (Fig. 2). The highly skewed sex ratio is readily apparent (Fig. 3); males accounted for less than 12% of the poeciliids (Palo Alto was exceptional with 18%). The correlation coefficient for the regression of percent female *P. mexicana* on distance downstream (excluding Santa Cruz and sites X and Z) was -0.793 ($p = 0.06$). Santa Cruz and sites X and Z were atypical. They were sampled only in 1977 and 1978, both years which yielded low frequencies of unisexuals throughout the tributary, and they consisted only of river habitat with no

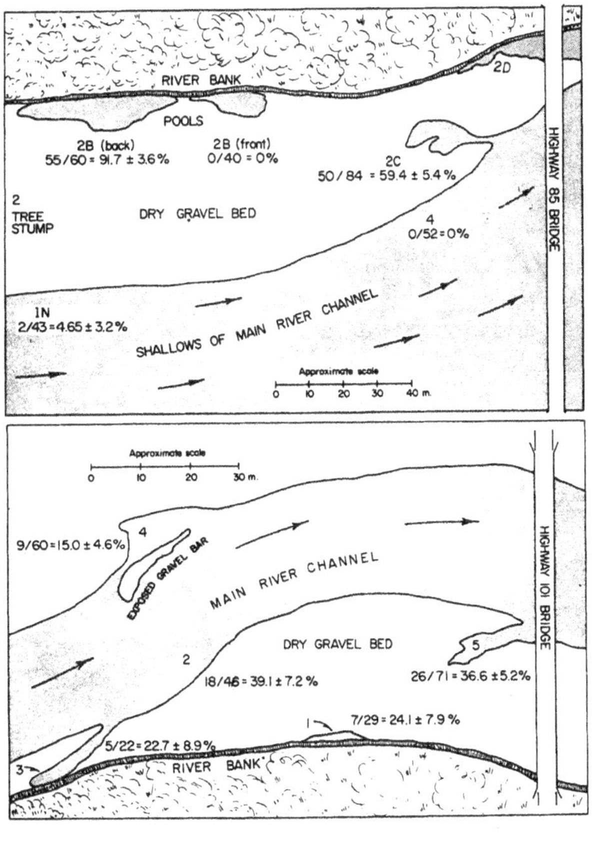

Fig. 4. Variation in the relative frequencies of triploid unisexuals collected in April, 1974 from various microhabitats in the Rio Purificacion at Barretal (upper) and Nuevo Padilla (lower). The relative frequencies (% ± 1SE) of the diploid unisexuals of *P. formosa* and the females of the bisexual *P. mexicana* are as follows:

Sites at Nuevo Padilla

	1	2	3	4	5
P. formosa	65.5 ± 8.8%	41.3 ± 7.3%	63.6 ± 10.3%	25.0 ± 5.6%	59.2 ± 7.1%
P. mexicana	10.3 ± 5.7%	19.6 ± 5.9%	13.6 ±7.3%	60.0 ± 6.3%	4.2 ± 2.4%

Sites at Barretal

	IN	2B(Back)	2B(Front)	2C	4
P. formosa	7.0 ± 3.9%	5.0 ± 2.8%	juveniles	26.4 ± 4.8%	5.8 ± 3.2%
P. mexicana	88.4 ± 4.9%	3.3 ± 2.3%	juveniles	14.3 ± 3.8%	94.2 ± 3.2%

backwaters. Santa Cruz had many diverse habitats due to the operation of a gravel pit.

The frequencies of the three types of females from different habitats in the same locality were compared to determine if there was microhabitat segregation (Fig. 4). The unisexual females were clearly more prevalent in the cul-de-sacs and backwater pools, whereas bisexual females of *P. mexicana* were more abundant in the shallows of the main river. The data support the hypothesis of microhabitat segregation among the three types of females, although both unisexual and bisexual females were usually found in a given habitat.

We predicted that unisexuals would prefer sandy substrate since they are found over sandy or silty substrates in the field. Our experiments (Fig. 5) yielded exactly the opposite result; bisexual female *P. mexicana* definitely preferred the sandy half of the large pool, but showed no preference between sunny (zone 1) and shady (zone 2) areas. Both unisexuals preferred the gravel substratum in the sun (zone 4) slightly more than the shaded area (zone 3). When this experiment was repeated in the field using an isolated pool (Table 1), the totals clearly indicated that all fish preferred the shady, gravel zones. The frequency in zone 4 with sunny gravel was the highest of the sunny zones because fish would momentarily swim into this zone from the two adjacent shady zones. Since the fish tended to stay together in the field experiment, a second set of habitat preference experiments was conducted to determine what would happen to a reduced number (i.e. five) of fish of each kind in large laboratory pools (Fig. 6). All three types of females were equally distributed over all four zones with only sand substrate, all three types were mostly in the

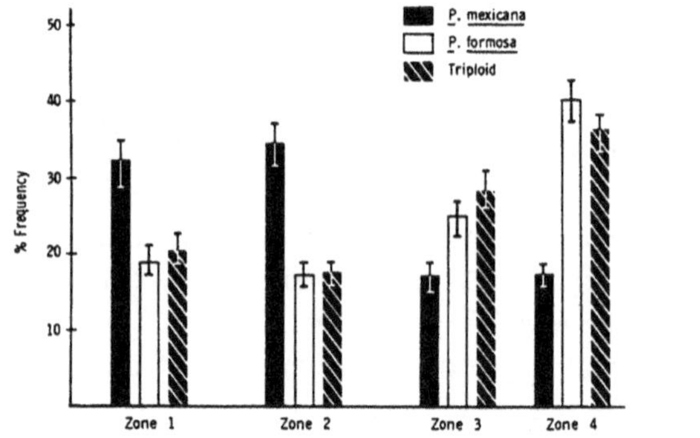

Fig. 5. Relative distribution of *Poecilia* females in four microhabitats within a large laboratory pool that contained ten of each kind of female. Zone 1: sunny with sand substratum; Zone 2: shaded with sand; Zone 3: shaded with gravel; Zone 4: sunny with gravel. Vertical lines at the top of each bar represent the 95% confidence limits for these percentages.

shade over an all gravel substrate, and when all four habitats were present, females were most often in the shade regardless of substrate.

Nearest neighbour data from field experiments show (Table 2) that females tend to be closer to their own kind than to different females ($p<0.001$, binomial test). Similar results were obtained from a pool with only *P. formosa* and *P. mexicana* present. Analysis of mean distances between females within radii of one, two or three standard lengths (Fig. 7) showed no significant differences for any comparison. Females appeared to associate in subgroups of two or three of their own kind within larger groups that might include all three types of females.

Since conspecific females outnumbered males, we had hypothesized that field caught *P. mexicana* males should prefer their own females over the unisexuals (McKay 1970, 1971), since they were sexually mature when collected and had been maintained in community tanks. Our first mate prefer-

Table 1. Field observation of habitat preference in *Poecilia*. Summary of 16 observations for 19 total hours of isolated pool with 10 females of *P. mexicana*, 6 of *P. formosa*, 8 triploids, and 3 males.

Zone:	3	1	4	2	6	8	7	5	
Habitat:	Shaded		. . . .	Sunny					
	. . . .	Rocks and gravel						Sandy	
Fish . .	Total number observed in each zone								Totals
P. mexicana	198	130	83	47	41	47	34	27	607
P. formosa	49	23	41	26	30	12	9	15	205
Triploid	194	114	124	84	68	62	58	52	756
Males	90	59	29	7	17	21	16	13	252
Zone totals	531	326	277	164	156	142	117	107	1820

Table 2. Summary of nearest neighbour field data. Based upon 39 observations for a total of 10 hours of isolated pool with 20 of each female in March, 1978.

Reference Fish	Number of times female was nearest neighbour *P. mexicana*	*P. formosa*	Triploid
P. mexicana	210	61	85
P. formosa	70	173	65
Triploid	91	68	173

ence experiments supported this prediction (Table 3), but subsequent experiments did not. These data were taken from an extensive behavioral study and included here to illustrate the variety of male responses. Most females were larger than males, and some bisexual females aggressively opposed the advances of males. Unisexual females appeared less aggresive. Although dominance hierarchies existed among males, mating frenzies that involved as many as 10 males attempting to inseminate one receptive female were observed in the field. So it seems that despite the skewed sex ratio, males were not in short supply. Furthermore, males were indiscriminate in their choice of mates, i.e. any receptive female. Cytological data indicate that even the smallest males had mature sperm throughout the year (Monaco et al. 1981), whereas the three types of females at any given site were asynchronous in egg development (Monaco et al. 1978).

Additional evidence that males are indiscriminate in their choice of mates (Table 4) comes from reproductive data for three types of females at four different series of collections. No significant differ-

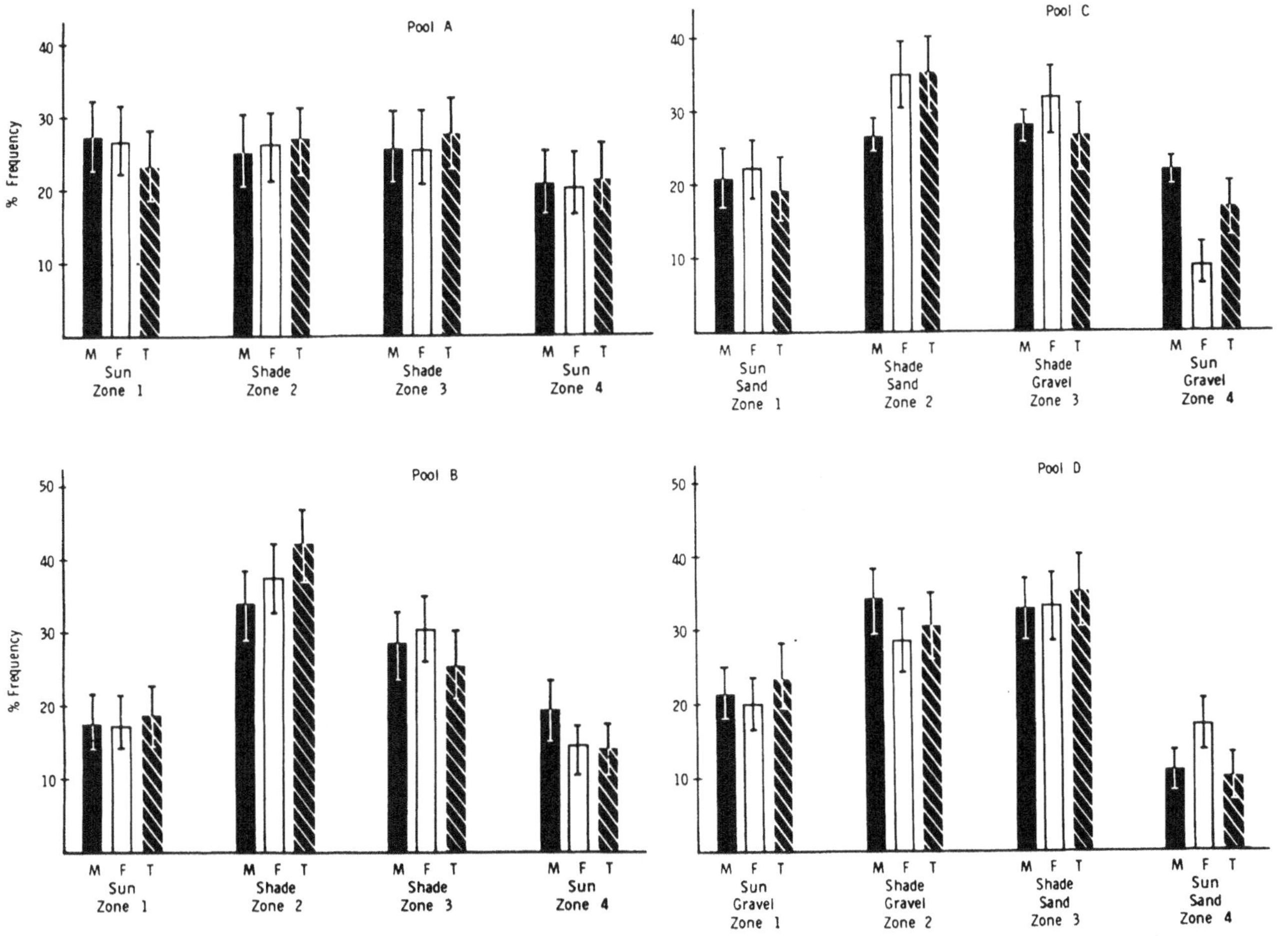

Fig. 6. Relative distribution of *Poecilia* females in different microhabitats within large laboratory pools that contained five of each kind of female. Pool A had all sand substratum whereas Pool B had only gravel; Pools C and D had all four microhabitats, but the arrangement of the four substrates in C was reversed for D. M: bisexual females of *P. mexicana*; F: unisexual females of *P. formosa*; T: triploid unisexuals. Vertical lines are the same as in Fig. 5.

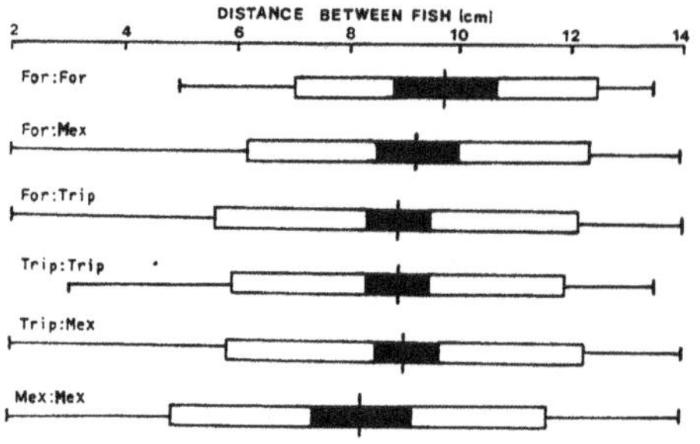

Fig. 7. Mean distances between paired females of *Poecilia*. Measurements were taken from projected slides taken in the field of 20 schools of tagged females observed at different times in an isolated pool containing 20 of each type of female. The distances between only those fish within a 14 cm radius i.e., twice the standard length of the reference fish, were measured. The horizontal line represents the total range of measurements, the open bar represents the mean plus and minus one standard deviation, and the black bar represents the 95% confidence interval of the observed mean which is shown as a vertical line.

ences existed in 1970 or 1971 collections between the proportion of pregnant bisexual and unisexual females. In 1972, however, there were significant differences. More unisexuals but fewer bisexuals were found pregnant in spring collections than expected, in the July collection more bisexuals and diploid unisexuals were found pregnant than expected, whereas fewer than expected triploids were pregnant.

There are no significant differences between bisexuals and unisexuals in the regression of number of embryos in eggs on standard body length. The bottom portion of Table 4 provides an index of relative reproductive output per female. There was little difference in mean number of embryos per female, but spring collection of *P. mexicana* females had a few more embryos and summer collection of *P. formosa* females had slightly more embryos. The triploids were consistenly lower than the other two types of females, perhaps their peak occurs in May (a possibility we are currently investigating).

Discussion

The reproductive success of the unisexuals indicated that they compete successfully with bisexual females for males, despite the greatly skewed sex ratio. The annual reproductive output per female may not differ significantly among different types, but times of peak reproduction do. Comparable data have been reported for *P. latipinna* and *P. formosa* populations near Brownsville, Texas (Hubbs 1964). That situation was also characterized by a highly skewed sex ratio, with a low, genetically-determined proportion of males, comparable fertilities in bisexuals and unisexuals but with different seasonal peaks, and wide fluctuations in fertility in females during the breeding season.

We think the asynchronous reproduction we observed, and that reported by Hubbs (1964), is a key feature allowing these species to coexist. The asynchrony also is in agreement with previous studies on ovarian cycles in these species (Monaco et al. 1978). Maturity of males appears to be independent of season, for a given body size (25 – 55 mm standard length) (Monaco et al. 1981).

Table 3. Mate preference of one male of *P. mexicana* with two females.

Male	Courting time with each female ($\bar{X} \pm$ SE in sec per 5 min period)			Number of gonopodial thrusts at each female ($\bar{X} \pm$ SE per 5 min observation period)		
	P. mexicana	*P. formosa*	Triploids	*P. mexicana*	*P. formosa*	Triploids
1	11.92 ± 1.03		7.81 ± 0.92**	11.88 ± 1.03		9.00 ± 1.56
1		8.06 ± 0.69	11.92 ± 7.47*		11.50 ± 0.85	9.25 ± 1.15
2	3.62 ± 0.24	5.37 ± 0.40**		2.10 ± 0.91	4.38 ± 0.98	
2		5.99 ± 0.91	4.35 ± 0.37		2.50 ± 1.28	2.50 ± 0.63
3	7.22 ± 0.63	7.85 ± 0.74	7.17 ± 0.82]*	5.80 ± 1.75	9.13 ± 0.52	3.60 ± 1.42]**
3			10.22 ± 0.72			12.50 ± 1.66
4	9.32 ± 0.93	9.98 ± 0.78		6.25 ± 0.90	12.00 & 2.16*	
4	10.85 ± 0.80		9.30 ± 0.66	13.13 ± 0.90		13.30 ± 0.73

* significant difference at 0.05 level
** significant difference at 0.01 level

Table 4. Relative reproductive output of *Poecilia* females in spring and summer collections

		P. mexicana		*P. formosa*		triploid	
		N	%	*N*	%	*N*	%
Frequency of females	Spring	837	49.5	40	67.5	152	52.0
with embryos in eggs:	Summer	863	29.3	200	28.0	446	28.7
Average number of	Spring	53	15.3	8	9.9	29	13.7
embryos in eggs per female:	Summer	66	20.8	32	31.9	32	16.9
Embryos produced	Spring	7.6		6.7		7.1	
per female:	Summer	6.1		8.9		4.9	
Relative reproductive	Spring	1.00		0.88		0.94	
Output:	Summer	0.68		1.00		0.55	

The lack of mate discrimination we observed in males is presumably reinforced by the fact that unreceptive bisexual females are more aggressive to males than are unreceptive unisexuals. Hubbs (1946), however, reported that male *P. latipinna* discriminate in favor of conspecific females, even allopatric males discriminated against unisexual females (but less so than sympatric males). McKay (1971) reported similar results for another bisexual-unisexual poeciliid complex. Bisexual male *Poeciliopsis lucida* show a marked preference for conspecifics and discriminated against both sympatric unisexual forms. Dominance hierarchies among males, plus a lack of experience in young males were postulated (McKay 1970, 1971) as the mechanisms responsible for the maintenaince of the unisexuals.

Moore & McKay (1971) estimated that a *P. lucida* male would inseminate an average of 1/100 unisexual female per 11-day period. How can we reconcile the lack of discrimination by *P. mexicana* males with the high degree of mate preference shown by *P. latipinna* and *P. lucida* males? One clue is readily apparent. The *Poeciliopsis* populations typically have a larger proportion of unisexual females (McKay 1971), as do the *Poecilia* populations from Texas (Hubbs 1964). The exact opposite is true in the populations we studied, where unisexuals typically comprise 20 – 40% of the population. Moore & McKay (1971) reported that over 90% of the unisexual *Poeciliopsis* females were pregnant when the female population was at least 22% bisexual. Moore (1976) estimated that at very low bisexual frequencies (i.e. less than 20%), the probability of unisexual pregnancy was 0.3 or less, whereas it was almost 1.0 for the bisexuals.

Since males comprise about 10% of the total *Poecilia* population at a given locality, and since females are receptive to males for only a few days around parturition (every 3 to 4 weeks from March to October), and since egg development and reproductive output are asynchronous for all three types of females, we infer that there are more males available to inseminate females on any given day than there are receptive females. The indiscriminant courtship of males in such circumstances agrees with the summarization of Hubbs (1946, p 67). The fact that females of all three types can carry viable sperm in their reproductive tracts throughout the year, and can also carry offspring in various stages of development (superfetation) would further reduce any competition for sperm.

Competition between unisexuals and bisexuals for space appears to be reduced by habitat segregation, both within and between habitats. Bisexuals preferred the shallows of the main river or headwaters, unisexuals preferred downstream areas, with more diversity. All females appeared to prefer to be in groups, even when 'foreign' females composed the group. Nearest neighbours were usually of the same type, so there were small subgroups of two or three similar females in larger groups containing all three types of females.

From the experiments described here, we infer at least three mechanisms which have evolved to reduce competition among the three types of females. Each type uses the services of males at different times, they use different segments of the river, and they use different microhabitats within a given segment of the river.

Acknowledgements

We thank the Mexican government for permission to carry out the field aspects of this research (number 2858 for 1977 was our most recent permit). Supported by grants from the National Science Foundation (most recently DEB 76-83184 to JSB and DEB 77-03257 and DEB 78-07753 to EMR), from the National Geographic Society to JSB & EMR, and from the University of Wisconsin-Parkside Committee on Research and Creative Activity to JSB.

References cited

Balsano, J.S., R.M. Darnell & P. Abramoff. 1972. Electrophoretic evidence of triploidy associated with populations of the gynogenetic teleost *Poecilia formosa*. Copeia 1972: 292–297.

Hardin, G. 1960. The competitive exclusion principle. Science 131: 1292–1297.

Hutchinson, G.E. 1965. The ecological theatre and the evolutionary play. Yale Univ. Press, New Haven. 139 pp.

Hubbs, C. 1964. Interactions between a bisexual fish species and its gynogenetic sexual parasite. Bull. Texas Mem. Mus. 8: 1–72.

Kucharski, K. 1979. Habitat preferences and associations of a unisexual-bisexual complex of *Poecilia* (Pisces: Poeciliidae) from northeastern Mexico. M.Sc. thesis, Florida Atlantic University, Boca Raton. 77 pp.

McKay, F.E. 1970. Behavioral aspects of population dynamics in unisexual-bisexual *Poeciliopsis* (Pisces: Poeciliidae). Ph.D. thesis, Univ. Connecticut, New Haven, 92 pp.

McKay, F.E. 1971. Behavioral aspects of population dynamics in unisexual-bisexual *Poeciliopsis* (Pisces: Poeciliidae). Ecol. 52: 778–798.

Moore, W. 1976. Components of fitness in the unisexual fish *Poeciliopsis monacha-occidentalis*. Evol. 30: 564–578.

Moore, W. & F.E. McKay. 1971. Coexistence in the unisexual-bisexual species complexes of *Poeciliopsis* (Pisces: Poeciliidae). Ecol. 52: 791–799.

Monaco, P.J., E.M. Rasch & J.S. Balsano. 1978. Cytological evidence for temporal differences during the asynchronous ovarian maturation of bisexual and unisexual fishes of the genus *Poecilia*. J. Fish Biol. 13: 33–44.

Monaco, P.J., E.M. Rasch & J.S. Balsano. 1981. Sperm availability in naturally occurring bisexual-unisexual breeding complexes involving *Poecilia mexicana* and the gynogenetic teleost, *Poecilia formosa*. Env. Biol. Fish. 6 (in print).

Rasch, E.M. & J.S. Balsano. 1974. Biochemical and cytogenetic studies of *Poecilia* from eastern Mexico. II. Frequency, perpetuation, and probable origin of triploid genomes in females associated with *Poecilia formosa*. Rev. Biol. Trop. 21: 351–381.

Env. Biol. Fish. Vol. 6, No. 1, pp. 49–58, 1981

Synchronized maturation and breeding in natural populations of *Xiphophorus variatus* (Poeciliidae)

Richard Borowsky & John Diffley

Department of Biology, New York University, Washington Square, New York, N.Y. 10003, U.S.A.

Keywords:
Social inhibition, Sexual development, Adaptive demography, Ovarian cycle, Fish, Platy

Synopsis

Reproductive events tend to be synchronized in natural populations of *Xiphophorus variatus*. Males physically resembling one another tend to mature at the same rate and time. This coordinate development has also been seen in the laboratory. We hypothesize that juveniles resembling each other experience fewer aggressive responses from males than do juveniles without 'look-alikes', because of habituation of aggression. This would decrease social inhibition and permit their synchronous development.

Females have an ovarian cycle. In the laboratory, crowding stress causes a suspension of breeding while a return to uncrowded conditions permits its resumption. Females released simultaneously from crowding tend to be synchronized in subsequent broods, although the synchrony deteriorates with time. Synchronizing events are shown to occur in natural populations.

The consequences of reproductive synchrony in males and females are briefly discussed.

Introduction

Poeciliid fishes inhabit the freshwater streams and rivers of Middle America under a wide variety of environmental conditions and have proven useful for the study of adaptive demography. Reproductive parameters differ among and within poeciliid species (Borowsky 1973a, Stearns 1977) as either genetic adaptation to local conditions or phenotypic responses to changes in environmental conditions (Stearns 1977, 1978, Borowsky 1978b, Thibault & Schultz 1978, Constanz 1979).

We have observed that reproductive events tend to be coordinated within natural populations of the variable platyfish, *Xiphophorus variatus*. Specifically, males of the same size and genetically determined tailspot phenotype tend to mature in concert, and females in the same pool tend to have their ovarian cycles in phase or exhibit evidence of synchronization prior to collection. Since both phenomena have potential stock size consequences, we document them here, present preliminary experimental data bearing on the mechanism of female reproductive synchrony, and briefly discuss possible consequences of reproductive synchrony.

Received 16.2.1980 Accepted 29.8.1980

Materials and methods

Fish collected in the field were preserved in 10% formalin from licalities given by Borowsky (1978a). Males were identified as to development of the anal fin which is undifferentiated in juveniles but mod-

© Dr. W. Junk b.v. Publishers, The Hague. Printed in The Netherlands. 0378-1909/81/061-049 $02.00.

ified as a sperm transfer organ (gonopodium) in adults. Transformation of the anal fin takes two to five weeks. We used a modification of the 'staging' system of Kallman & Schreibman (1973); we subdivided their stage 2 more finely by counting the number of divisions in the third fin ray (Borowsky 1973a). Developmental stages of young were modified (Fig. 1) from Tavolga (1949).

The tailspot system of *X. variatus* is under the control of a series of autosomal dominant alleles (Borowsky 1978a). In studied stocks there are two spotting alleles, *C* and *Ct*, the former producing the crescent pattern, the latter the cut-crescent pattern. Heterozygotes (*C/Ct*) have both patterns. Wild-type fish (+/+) lack tailspots.

Maturation of male *X. variatus* is subject to social inhibition. The larger of two males will inhibit the smaller one, but not vice versa. The larger one therefore matures first and consequently stops growing (Borowsky 1973a). The inhibited male continues to grow until it nearly attains the size of the first male and then commences maturation (Borowsky & Scelta unpublished). Since growth continues during the maturation process, the second male generally grows to be larger than the first (Borowsky 1973a). In small groups (4 to 12 fish) sexual development usually proceeds in size order, largest first (Borowsky 1973b).

Ocassionally, two or more males commence development at the same time and proceed at the same rate. The high degree of coordination (Fig. 2) suggests a synchronizing mechanism. This coordinated development has been observed in no more than six of 46 pairs. It becomes more common in

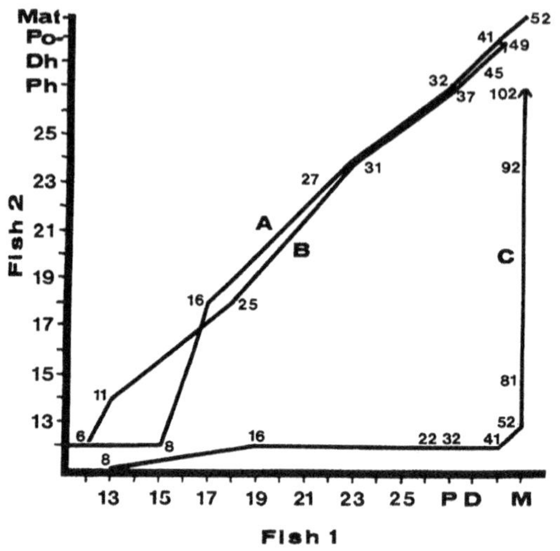

Fig. 2. Coordinate development in *X. variatus*. Maturational histories of three pairs of fish are figured. The developmental stages of the second male to mature are plotted against those of the first. A and B illustrate coordinate development. C illustrates a more typical pair in which one fish matures before the other. The numbers on the curves denote days from the start of observations. Stages of development are modified from Kallman & Schreibman (1973; Ph, Dh, Po- and Mat correspond to stages 3 to 6 in their system. Numbers along the axes denote the number of divisions in the third ray of the anal fin during stage 2.

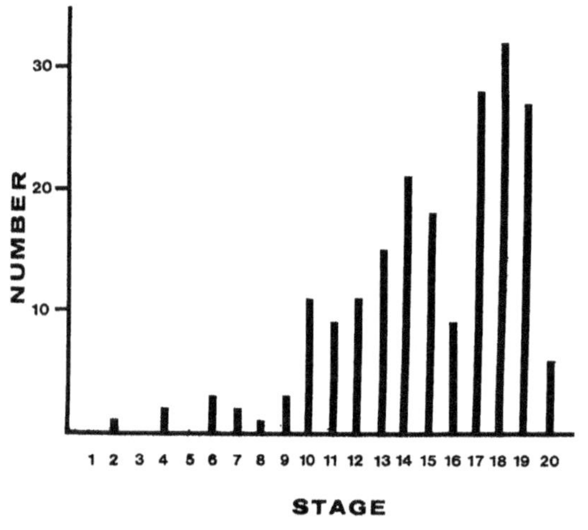

Fig. 1. The distribution of broods among developmental stages for wild-caught females. Stages modified after Tavolga (1949) as follows: stage 1 = Tavolga's stages 1 and 2, 2 = 3 and 4, 9 = 11 and 12, 16 = 19 and 20, 17 = 21 and 22, all other stages correspond in sequence.

larger groups (Fig. 3), but remains the exception in the laboratory.

In natural stocks, however, the synchronization is more common. Sexually maturing males of identical tailspot phenotype and similar length tend to be at the same stage of development within collections. The 12 largest collections were subjected to a nearest-neighbour analysis as follows: developmental stage was plotted against size for all maturing males in a given pool (Fig. 4). The axes were then standardized so that total variation along the abcissa equalled that along the ordinate, and variation along each axis was equivalent. Each point (= fish) within the plane was then identified according to tailspot phenotype and its nearest neighbour determined as like or unlike phenotype. The expected frequency of nearest neighbour being 'like' is:

$$\text{Exp (Like)} = \sum_{i=1}^{t} \frac{n_i (n_i - 1)}{N - 1},$$

where t is the number of the phenotypes, n_i the frequency of the ith phenotype, and the total number of individuals is N.

Female *Xiphophorus* have a regular ovarian cycle (Siciliano 1972). Brood intervals vary with temperature, and superfetation is rarely seen (Kallman 1975). In the laboratory at 22 ± 2° C, brood intervals for *X. variatus* vary among lines but average 42 – 45 days (Fig. 5, Borowsky unpublished). Females over 22 mm standard length are capable of breeding and normally cycle independently of each other. Minimum size at reproduction varies among collections from the same locality, with the smallest reproducing females at Encino and Sarco 20.2 and 21.2 mm, respectively.

Results

Sexual maturation in males

Observed coordinate development in males was compared to expected, summed over all collections and tested for significance by Chi-square analysis (Table 1). In 11 of 12 collections, the observed frequencies exceeded expected, and overall this discrepancy was highly significant ($X^2 = 26.5$, df = 1, $p < 0.0001$), so we accept the hypothesis that like phenotypes tend to be clustered. The biological interpretation of these clusters is that they represent sets of males exhibiting coordinated development.

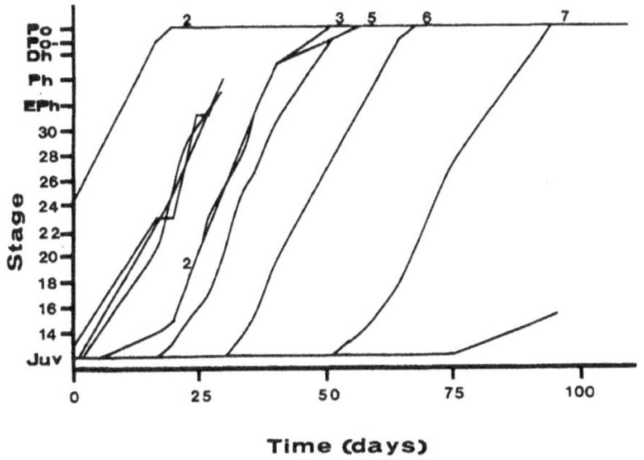

Fig. 3. Coordinate development in a group of males. Developmental stages are plotted against time for eleven males. Each line represents a single male unless numbered otherwise. Three groups of fish exhibit coordinate development. Eph = early Ph stage.

Synchronized breeding of females

A total of 456 females over 20 mm length from 20 collections made at seven localities were dissected. Of these, 33% had immature ovaries, 24% contained developing eggs, and 44% contained young in various stages of development (Fig. 1). In most collections there was a positive correlation between female size and egg or embryo number. However, repeated samples from the same stock, even from adjacent pools on the same day, exhibited highly

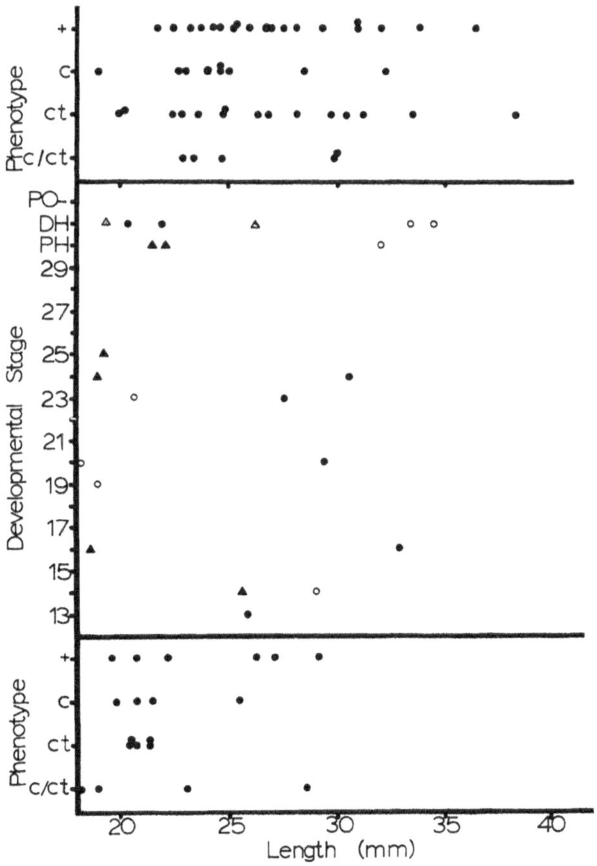

Fig. 4. Lengths of males in a natural population (Sarco, pool 1, 1973, see Table 1). Each point represents a single fish. The upper and lower diagrams give the size distributions of mature and immature males, respectively (+ = wild-type; C = crescent; Ct = cut-crescent; C/Ct = crescent/cut-crescent heterozygotes). The middle portion plots stage of development against length for maturing males (open circles = +; triangles = C; dots = Ct; open triangles = C/Ct). Maturing males cluster by phenotype. Males mature within the size distribution of adults.

significant differences among regressions (Table 2, Fig. 6).

Field evidence suggested that traumatic conditions can suspend reproduction in females. For example, a large collection made at Arroyo Sarco in April 1975, at the end of the dry season was in warm (24° C), stagnant water. Fish were crowded and in poor condition, and individuals of less-hardy species (*Cichlasoma steindachneri*, *Astyanax mexicanus*) were dead and dying. Eighty females of reproductive length (mean length 32.6 mm) showed 66 with regressed ovaries, four with developing eggs and ten with developing embryos in advanced stages. It was unusual for so many females not to be cycling, and for the few with developing embryos to be in similar stages.

Table 1. Nearest neighbor analysis to test for clustering of like male phenotypes on developmental plane.

Collection		*Phenotype of nearest neighbor* Like	Unlike
Arroyo Sarco			
Pool 6, 1967	Observed	15	10
	Expected	9.25	15.75
Pool 1, 1973	Observed	12	10
	Expected	5.52	16.48
Pool 6, 1973	Observed	7	14
	Expected	4.9	16.1
Pool 1, 1975	Observed	13	11
	Expected	9.3	14.7
Pool 1, 1977	Observed	8	8
	Expected	6.8	9.2
Pool 6 (Lower), 1977	Observed	4	8
	Expected	4.55	7.45
Pool 6 (Upper), 1977	Observed	7.5 (Tied Obs.)	13.5
	Expected	7	14
Arroyo Encino			
Big Pool, 1970	Observed	2	4
	Expected	1.17	4.83
Big Pool, 1973	Observed	4	4
	Expected	2.07	5.93
Upper Pool, 1975	Observed	10	1
	Expected	5.4	5.6
Pequeseh			
1977	Observed	14	1
	Expected	9.86	5.14
Aquismon			
1977	Observed	9	2
	Expected	4.6	6.4
Summed observations		105.5	90.5
Summed expectations		70.42	125.58
Continuity corrected $X^2 = 26.5$, df = 1, $P = 2.6 \times 10^{-7}$			

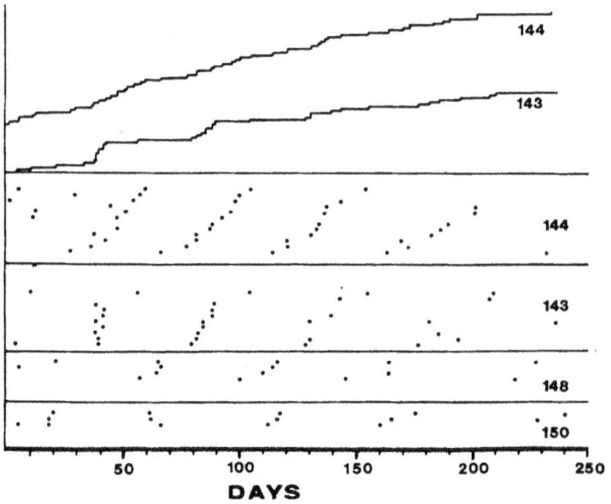

Fig. 5. Brood records for four groups of wild-caught females (143, 144, 148 and 150). Each dot represents a single brood and all broods of the same female are plotted on the same horizontal line. Time is measured from the date (April 24, 1975) of the first brood in group 144. Group 143 was collected on April 11 and group 144 on April 17; females from both groups were placed in individual aquaria on April 21 to April 23. Cumulative brood records for both groups are plotted at the top and reveal better synchrony in group 143 than in group 144.

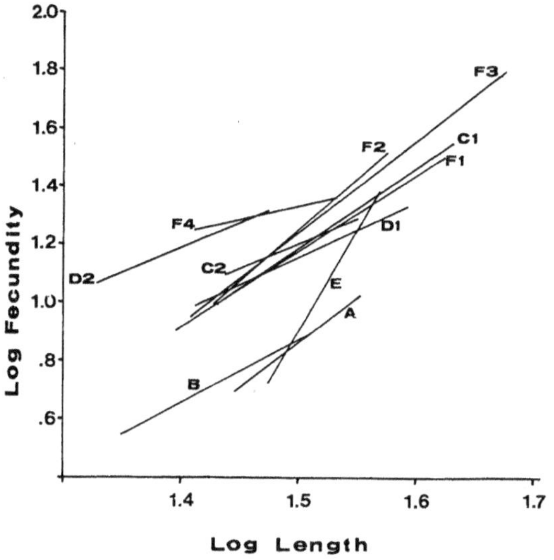

Fig. 6. The relationship between length and fecundity in natural populations varies significantly among collections (Table 2). Lines of best fit for the regression of log number of mature eggs or developing embryos against log female length (mm) are plotted for eleven collections from two localities. Encino: A = 1970; B = 1975; C1 and C2 = collections from adjacent pools in 1977. Sarco: D1 and D2 = pools 1 and 6, 1973; E = pool 1, 1975; F1 to F4 = pools 1, 3, 6 (lower), and 6 (upper), 1977.

Table 2. ANCOVA results for tests of significant differences among size-fecundity relationships in wild-caught females. F ratios and significant probabilities for differences in slopes and elevations among selected collections. Letter refer to regression lines plotted in Figure 6.

Collections compared	Slope	Elevation
Arroyo Sarco		
All seven collections	$F_{6,152} = 2.02$	$F_{6,158} = 15.2$
(D through F)	$P = 0.07$	$P = 10^{-13}$
Pool 1 *vs* Pool 6, 1973	$F_{1,32} = 0.36$	$F_{1,33} = 16.2$
(D1 *vs* D2)		$P = 3 \times 10^{-4}$
All four 1977 collections	$F_{3,108} = 1.73$	$F_{3,111} = 9.3$
(F)	$P = 0.16$	$P = 10^{-5}$
Pool 6 Lower *vs* Upper	$F_{1,52} = 6.68$	$F_{1,53} = 17.45$
1977 (F3 *vs* F4)	$P = 0.013$	$P = 10^{-4}$
Arroyo Encino		
All four collections	$F_{3,35} = 0.27$	$F_{3,38} = 15.34$
(A through C)		$P = 10^{-6}$

If poor conditions can cause a suspension of breeding, and return to good conditions trigger its resumption, a group of females subjected to the same, sudden environmental change should be synchronized. This 'stress and release' hypothesis was first suggested to us by the breeding performance of mature females collected in nature, stored in the field under crowded conditions for five to 15 days, and then transported back to the laboratory where they were placed individually in aquaria. These females had been inseminated in nature and produced from one to six broods from stored sperm. Their reproductive records exhibited a high degree of synchrony (Fig. 5) which decreased in subsequent broods as females moved out of phase. This was the key to detecting stress and release effects in natural populations.

We tested the stress-release hypothesis in the laboratory by taking mature females from community tanks at low densities (about 0.25 fish l^{-1}), transferring them to more crowded tanks (stress) for periods of three to eleven days, then transferring them to individual tanks (release) (Table 3). The three day crowding period in experiment I failed to synchronize the females, but clearly affected other aspects of their reproduction (Table 4). Significantly fewer young were produced in first broods after crowding ($N = 13$, $\bar{x} = 7.9 \pm 4.9$) than in second broods ($N = 11$, $\bar{x} = 20.5 \pm 8.7$, $t_{22} = 4.5$, $p = 2 \times 10^{-4}$). The mean brood size for second broods is typical for the species, the difference can be attributed to the treatment.

The mean brood interval in experiment I was about 45 days, with seven of 13 first broods born within 34 days of crowding. They were therefore caught in development, rather than prior to fertilization (Tavolga 1949). The missing young of these females must have been aborted and resorbed. The remaining six females probably were treated prior to fertilization and their reduced brood size could reflect a reduced number of eggs fertilized. Susceptibility to abortion appears to decline with development. Among the seven females giving birth within 34 days, brood size was negatively correlated with birth date ($r = -0.078$, two-tailed $p = 0.04$).

The effects of the treatment were more striking in experiment II. We distinguished between first interval broods (born within 34 days of treatment) and second interval broods (first broods born 53 or more days after treatment, or second broods). There was again a negative correlation between brood size and birth date for broods born on or before day 33, but the correlation was not itself significant ($r = -0.26$). When the results of both experiments were evaluated by multiple regression, an overall significant trend emerged ($F_{2,21} = 4.4$, $p = 0.05$). No broods were born during the interval between days 33 and 46. To have given birth during that interval females would have had unfertilized, developing eggs during the treatment. Our interpretation is that fertilization did not occur in those females. First broods

Table 3. Treatments in the Stress and Release experiments. The females used were the F_1 and F_2 of fish collected in 1976 from the Rios Nautla and Axtla in Mexico. All females were transferred to individual, eight liter aquaria after crowding.

Experiment	Pedigrees and origin	Crowding treatment
SRI	N243 (Nautla)	*3 days*, 13 fish per liter. (N = 13)
SRII	N243, N250, N254, N270 (Nautla), N247 (Axtla)	*3 days*, 3.6 fish per liter. Treatment A (N = 12) *6 days*, 3 days at 3.6 fish per liter, 3 days at 2.5 fish per liter. Treatment B (N = 12) *11 days*, 3 days at 3.6 fish per liter, 3 days at 2.5 fish per liter, 5 days at 1.4 fish per liter. Treatment C (N = 15)

averaged 3.9 (± 2.9) young (N = 17), while second intervals broods averaged 11.9 (± 6.9) young (N = 18, t_{33} = 4.4, p = 10^{-4}). Again, we conclude that the young 'missing' from first broods had been aborted and resorbed.

Nearly all young in experiment II were weighed to the nearest 0.1 mg within 12 h of birth. When broods were matched by females, young of second broods were smaller and less variable than those of first broods (p = 0.008, Table 4). Birth dates of second intervals broods were strongly correlated with treatment (Fig. 7). For the three-day crowded females the mean birth date for second interval broods was 56.2 days after crowding, for six-day females it was 60.2 and for eleven-day females it was 67.7 (r 0.999, t_1 32.7, p 0.01). The degree of synchrony within treatment in this experiment, while not as good as that in Figure 5, does show that effects of simultaneous return to good conditions can synchronize females.

In several field collections a highly significant correlation exists between female size and developmental stage of their young (Fig. 8, Table 5). Of the 17 largest collections, three had significant correlations (two-tailed, p = 0.01). When the probability values for all 17 collections were combined according to the method of Fisher (1946), the overall level of significance was p = 2.5 × 10^{-4}. A two-sided analysis was chosen because in some cases the observed correlation was positive and in others it was negative.

Discussion

Males

Our working hypothesis is that social inhibition, mediated through agonistic behavior, coordinates maturation of males. The more often a juvenile is the recipient of agonistic attention, the more strongly it will be inhibited from maturing. Agonistic behavior in fishes is subject to habituation (Clayton & Hinde 1968, Gallagher et al. 1972). A single or distinctive juvenile *X. variatus* might elicit much aggression from adult males and remain inhibited while one among a group of 'look-alikes' might elicit fewer such responses because of habituation. Therefore groups of juveniles closely resembling each other would be able to break through inhibition and mature in concert.

Social inhibition of maturation in natural stocks of *X. variatus* buffers stocks against environmental fluctuations affecting the availability of adult males (Borowsky 1978b). For example, selective predation on adult males would release juveniles from inhibition and increase recruitment rate to adults. This recruitment involves synchronized development of like juveniles, possibly through habituation of aggression. If correct, the synchrony may be a gratuitous consequence of the social inhibition system, rather than an adaptive phenomenon itself.

Nevertheless, coordinated sexual development of males has several potentially significant demogra-

Table 4. Mean weights of progeny in first and second broods, matched by female parent. Weight (mg) ± one standard deviation, brood size in parentheses.

Female	*First brood*	*Second brood*	*Student's t*	*Two-sided probability*
A	21.0 (1)	6.9 ± 1.2 (7)		
B	10.6 (1)	10.2 ± 1.3 (6)		
C	19.0 (1)	8.6 ± 0.2 (2)		
D	12.7 (1)	9.9 ± 1.0 (22)		
E	14.8 ± 3.1 (6)	9.8 ± 1.1 (13)	t_{17} = 5.3	6.1 × 10^{-5}
F	10.1 ± 1.1 (11)	9.0 ± 1.1 (16)	t_{25} = 2.6	1.7 × 10^{-2}
G	11.0 ± 1.2 (3)	9.8 ± 0.8 (20)	t_{21} = 2.3	3.3 × 10^{-2}
H	11.0 ± 0.8 (3)	10.6 ± 2.0 (19)	t_{20} = .34	7.4 × 10^{-1}

Summary statistics: First brood mean = 12.2 mg (N = 27)
Second brood mean = 9.7 mg (N = 105)
Overall X^2 = 35, df = 8, P = 3 × 10^{-5}

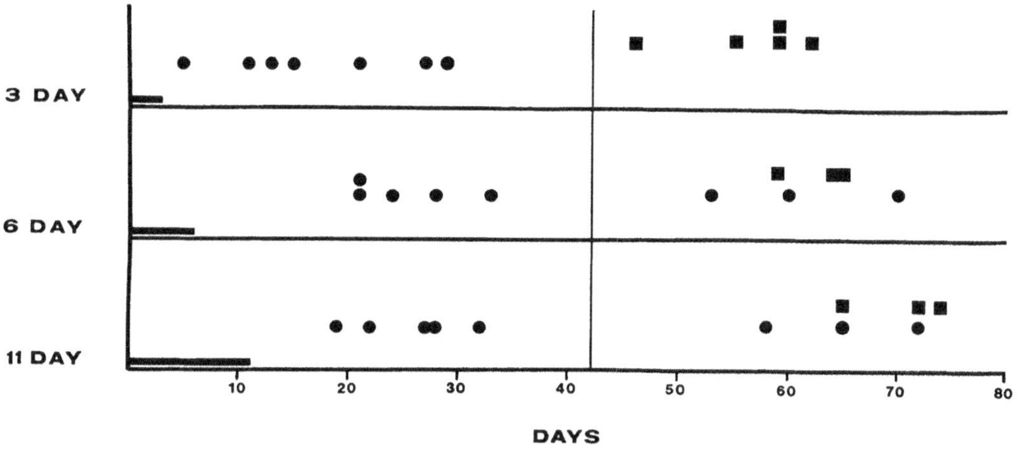

Fig. 7. Brood records for the females (treatments A, B, and C, cf. Table 3) in the second stress and release experiment. Time measured from the start of crowding with the thick horizontal lines marking the duration of treatment. Dots denote first broods and squares denote second broods. The vertical line at day 42 marks one full ovarian cycle. The timing of second interval broods is correlated with the end of treatment.

phic and population genetic consequences. The presence of other males of the same tailspot phenotype facilitates maturation, decreases age at maturity, and reduces generation time. The magnitude of this effect would be frequency-dependent (positive feedback) for intermediate morph frequencies; the greater the frequency of a morph, the greater the advantage. On the other hand, males that matured in concert would compete more for females with one another than with other males. The magnitude of this effect would also be frequency-dependent (negative feedback) and could cancel the positive effects. The importance of intramorph sexual competition would be increased if mature males have a short life expectancy in natural populations, as has been suggested by other investigators (Borowsky 1973b, Farr 1975, Borowsky & Kallman 1976, Endler 1978).

Our hypothesis assumes that inhibition is correlated with the amount of agonistic behavior directed towards a male. Sohn (1977a) suggested that dominant male *X. maculatus* are not inhibited while subordinates are, and that a dominance reversal by the second fish as it continues to grow is the means by which it overcomes inhibition. However, the observation that the first fish to mature is dominant adds little to the original observation that the first fish to mature is larger (Borowsky 1973a), because dominance is a correlate of size. The crucial test of the dominance hypothesis is the social status of the

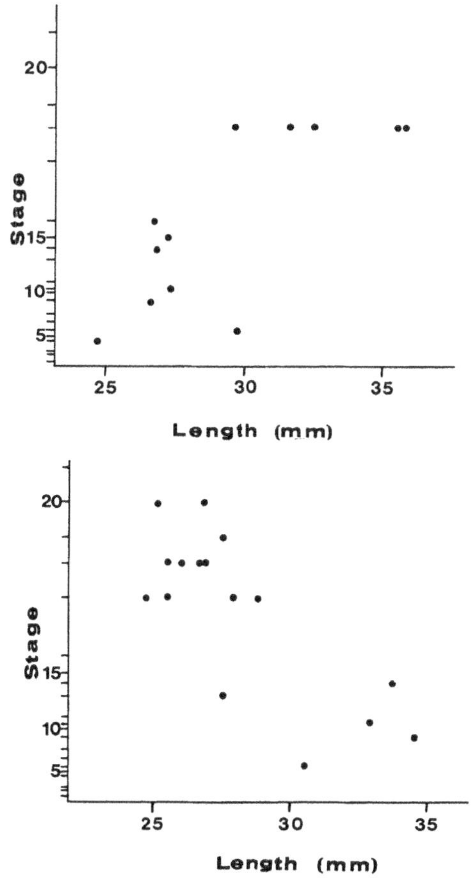

Fig. 8. Developmental stage of young plotted against length of the female parent. Top: Females collected January 1967 at Jaumave, above dam. Pearson's r = 0.74, two-sided P = 0.006. Below: Females collected in March 1977 from the Rio Sabinas. Pearson's r = 0.66, two-sided P = 0.005.

second male when it initiates maturation, and these data have not been reported by Sohn. We reject an unmodified dominance hypothesis since: 1) In natural stocks, males mature in the presence of larger males (Fig. 4, and Borowsky 1978b), while the dominance hypothesis predicts that maturing males should be the largest in the stock. 2) The degree of inhibition in natural stocks is correlated with the proportion of adults present (Borowsky 1978b). The dominance hypothesis does not predict that inhibition should be quantitative. 3) In 12 of 13 cases in the original data of Borowsky (1973a) the second fish had started to mature while still smaller than the first (Borowsky & Scelta unpublished). 4) Unpublished data on *X. maculatus* and *X. variatus* pairs show that second males usually initiate development and sometimes complete it while still subordinate. These observations are inconsistent with the dominance hypothesis but consistent with our tentative hypothesis.

Females

We propose that environmental fluctuations can initiate synchronous breeding of females in natural stocks when conditions improve. Maglio & Rosen (1969) reported that reproduction of *Gambusia affinis* in an artificial pond in New York was synchronized once a year, at the start of breeding in April, but by August and September females were cycling out of phase. In contrast, natural stocks of *X. variatus* could be synchronized several times during their breeding season if several substantial rains occurred.

Breeding synchrony in *X. variatus* females appears to be a gratuitous result of environmental fluctuations but has potentially important population consequences. Synchronized births would flood a locality with young and reduce the probability of their predation. Poeciliid females are most receptive to courtship in the first few days after parturition (Liley 1966) so mating activity in the stock would also tend to be synchronized. This might afford significant protection from predation to males by making them all conspicuous simultaneously. Breeding synchrony could decrease the importance of sexual selection. Gandolfi (1971) reported that when two *Poecilia reticulata* males are placed with a single female, the female is fertilized by the dominant male. When two females are placed with the males, however, the subordinant male sires some of the offspring. That is, sexual selection decreases in importance as females increase in availability.

Stream conditions in northeastern Mexico are highly variable during the dry season (Borowsky 1978a), and this variability significantly affects female fecundity (Fig. 6). Precipitation is seasonal in this region, with only occasional rains from September to April (Darnell 1962). These rains cause improvements in environmental conditions which could synchronize females by stimulating reproduction (Stearns 1977), as Figure 6 and Table 2 confirm for *X. variatus*. Differences in size-specific fecundity are almost as great among repeated samples from the same locality as among localities.

As environmental conditions degenerate at the end of the dry season, most of the reproductive females probably simply stop cycling. Developing young might be aborted or fertilization might not occur. Abortion appears not to be an important mechanism in natural stocks (only nine of 199 females collected with developing young contained

Table 5. The correlations between female length and developmental stages of progeny for seventeen collection from Mexico, 1967 to 1977.

Locality and date	Sample size	Pearson's correlation coefficient	Two sided probability
Arroyo Sarco			
Pool 1, 1973	8	0.68	0.063
Pool 1, 1976	6	0.055	0.96
Pool 1, 1977	34	−0.076	0.67
Pool 3, 1977	9	0.26	0.50
Pool 6 (Lower), 1977	22	−0.14	0.53
Pool 6 (Upper), 1977	7	0.83	0.020
Arroyo Encino			
Stream, 1973	6	−0.66	0.16
Enclosure, 1973	9	0.23	0.55
Upper pool, 1975	10	0.57	0.089
Jaumave			
Above dam, Jan. 1967	12	0.74	0.0059
Above dam, Feb. 1967	9	0.15	0.70
Below dam, Feb. 1967	7	0.048	0.92
Rio Frio, 1977	11	0.43	0.19
Rio Sabinas, 1977	16	−0.66	0.0051
Tanlajas, 1977	9	0.89	0.0012
San Joachin, 1977	6	−0.18	0.74
Pequeseh, 1977	6	−0.72	0.10

aborted progeny). In contrast, all females in the stress-release experiments that had young at the time of crowding had reduced first brood sizes. Since we place so much emphasis on the effects of suboptimal conditions in nature we must ask why abortions are so rarely observed in our collections? We believe the answer is related to the rate at which conditions deteriorate. In our experiments, conditions were altered rapidly, and for females with developing young, abortion was the only mechanism available. In the field, however, conditions may degenerate over days or weeks, and females may be able to reduce brood size by decreasing the number of eggs fertilized.

It appears that the significantly smaller first brood size and significantly larger young are related in that nutrients resorbed from aborted siblings are recycled to surviving embryos. Such an effect would clearly be adaptive because the young would be born larger and therefore more likely to survive. Anatomical structures permitting the mother to nourish developing embryos are well developed in poeciliids of the genera *Poeciliopsis* and *Heterandria*, but evidence for such effects in *Xiphophorus* is indirect (Scrimshaw 1944, 1946, Thibault & Schultz 1978).

We hypothesize that female size is a correlate of brood interval in natural stocks, and that the correlation between size and developmental stage arises as females move out of phase following a synchronizing event. Under some circumstances smaller females cycle more quickly than larger ones, and a negative correlation arises, while under others smaller females cycle more slowly than large ones and the correlation becomes positive. We suggest that when food is abundant, larger females cycle more quickly and when food is limited, smaller females cycle more rapidly. We have no data on this, but we are currently testing the hypothesis.

Poeciliids are being increasingly studied in the field to test theories of adaptive demography (Borowsky & Kallman 1976, Sohn 1977, Stearns 1978, Constanz 1979), but as our results show, demographic data should be interpreted cautiously and in terms of experimental work because of phenotypic plasticity.

Acknowledgements

We gratefully acknowledge support by NSF grants BMS 75-13464 and DEB 79-12341. Ms. J. Scelta provided technical assistance. Permission to collect was obtained from the Mexican authorities.

References cited

Borowsky, R.L. 1973a. Social control of adult size in males of *Xiphophorus variatus*. Nature 245: 332–335.

Borowsky, R.L. 1973b. Relative size and the development of fin coloration in *Xiphophorus variatus*. Physiol. Zool. 46: 22–28.

Borowsky, R.L. 1978a. The tailspot polymorphism of *Xiphophorus* (Pisces: Poeciliidae). Evol. 32: 886–893.

Borowsky, R.L. 1978b. Social inhibition of maturation in natural populations of *Xiphophorus variatus* (Pisces: Poeciliidae). Science 201: 933–935.

Borowsky, R.L. & K.D. Kallman. 1976. Patterns of mating in natural populations of *Xiphophorus* (Pisces: Poeciliidae). I: *X maculatus* from Belize and Mexico. Evol. 30: 693–706.

Clayton, F. & R.A. Hinde. 1968. The habituation and recovery of aggressive display in *Betta splendans*. Behaviour 30: 96–106.

Constantz, G.D. 1979. Life history patterns of a livebearing fish in contrasting environments. Oecol. 40: 189–201.

Darnell, R.M. 1962. Fishes of the Rio Tamesi and related coastal lagoons in East-Central Mexico. Publ. Inst. Mar. Sci. (Port Aransas, Texas) 8: 299–365.

Endler, J.A. 1978. A predator's view of animal color patterns. Evol. Biol. 11: 319–364.

Farr, J.A. 1975. The role of predation in the evolution of social behavior of natural populations of the guppy, *Poecilia reticulata* (Pisces: Poeciliidae). Evol. 29: 151–158.

Gallagher, J.E., M.J. Herz & H.V. Peeke. 1972. Habituation of aggression: the effects of visual social stimuli on behavior between adjacently territorial convinct cichlids (*Cichlasoma nigrofasciatum*). Behav. Biol. 7: 359–368.

Gandolfi, G. 1971. Sexual selection in relation to the social status of males in *Poecilia reticulata* (Teleostei: Poeciliidae). Boll. Zool. 38: 35–48.

Kallman, K.D. 1975. The platyfish, *Xiphophorus maculatus*. pp. 81–132. *In*: R.C. King (ed.) Handbook of Genetics, Vol. 4. Plenum Press, New York.

Kallman, K.D. & M.P. Schreibman. 1973. A sex-linked gene controlling gonadotrop differentiation and its significance in determining the age of sexual maturation and size of the platyfish, *Xiphophorus maculatus*. Gen. Comp. Endocrinol. 21: 287–304.

Liley, N.R. 1966. Ethological isolating mechanisms in four sympatric species of poeciliid fishes. Behaviour Suppl. 13: 1–197.

Maglio, V.J. & D.E. Rosen. 1969. Changing preference for substrate color by reproductively active mosquitofish, *Gambusia affinis* (Baird and Girard) (Poeciliidae, Atheriniformes). Amer. Mus. Novit. 2397: 1–37.

Scrimshaw, N.S. 1944. Embryonic growth in the viviparous poeciliid, *Heterandria formosa*. Biol. Bull. 87: 37–51.

Scrimshaw, N.S. 1946. Egg size in poeciliid fishes. Copeia 1946: 20–23.

Siciliano, M.J. 1972. Evidence for a spontaneous ovarian cycle in fishes of the genus *Xiphophorus*. Biol. Bull. 142: 480–488.

Sohn, J.J. 1977a. Socially induced inhibition of genetically determined maturation in the platyfish, *Xiphophorus maculatus*. Science 195: 199–201.

Sohn, J.J. 1977b. The consequences of predation and competition upon the demography of *Gambusia manni* (Pisces: Poeciliidae). Copeia 1977: 224–227.

Stearns, S.C. 1977. The evolution of life history traits: a critique of the theory and a review of the data. Ann. Rev. Ecol. Syst. 8: 145–171.

Stearns, S.C. 1978. Interpopulation differences in reproductive traits of *Neoheterandria tridentiger* (Pisces: Poeciliidae) in Panama. Copeia 1978: 188–190.

Tavolga, W.N. 1949. Embryonic development of the platyfish (*Platypoecilius*) and swordtail (*Xiphophorus*) and their hybrid. Bull. Amer. Mus. Nat. Hist. 94: 165–229.

Thibault, R. & R.J. Schultz. 1978. Reproductive adaptations among viviparous fishes (Cyprinodontiformes: Poeciliidae). Evol. 32: 320–333.

Env. Biol. Fish. Vol. 6, No. 1, pp. 59-64, 1981

Maintenance of female mimicry as a reproductive strategy in bluegill sunfish (*Lepomis macrochirus*)

Wallace J. Dominey

Field of Neurobiology and Behavior, Langmuir Laboratory, Cornell University, Ithaca, New York 14853, U.S.A.

Keywords:
Behavior, Reproduction, Sexual selection, Mating strategies, Pseudofemale, Coloniality, Centrarchidae

Synopsis

In species where male reproductive success is dependent on male competition and aggression, alternative reproductive patterns, thought to represent a reduction in male reproductive effort, sometimes occur. Female mimicry in bluegill sunfish (*Lepomis macrochirus*) is an example of an obligate alternative male strategy. Female mimics are small, sexually mature males which mimic the details of female behavior, and gain access to functional females attracted to the nests of large, aggressive territorial males. The costs of female mimicry relative to nesting male behavior are discussed and two hypotheses, deception and mutual gain, are presented to explain the tolerance and 'courtship' of female mimics by nesting males.

Introduction

In species where male reproductive success is dependent upon male-male competition and aggression, individuals have been observed to adopt wholly different, mutually exclusive constellations of reproductive behavior. These alternative reproductive patterns are often 'submissive' patterns, and are thought to reduce the costs associated with male-male competition, such as an increased risk of predation, time and energy loss, and an increased probability of injury due to fighting. In addition, males which adopt these low cost reproductive patterns are often disadvantaged in reproductive competition, perhaps due to their small size, lack of weapons, or limited experience. Ostensibly, then, alternative patterns occur when the costs of displaying typical male behavior are high, and the probability of successful competition by an individual is correspondingly low.

In cases where individuals practice both the typical and the alternative reproductive pattern opportunistically, or over the course of development, these reproductive options can be considered to be analogous to courtship signals. Each signal (or pattern) is displayed when most appropriate, and the incremental increase in fitness attributable to each signal need not be equal for both patterns to be expressed. In contrast, if individuals practice only a single reproductive option, the maintenance of two patterns in the population might indicate a balanced genetic polymorphism, where the costs and benefits of each reproductive pattern sum to give both patterns the same fitness value. Such a system, for the coevolution of a high cost and a low cost

Received 16.1.1980 Accepted 28.8.1980

© Dr. W. Junk b.v. Publishers, The Hague. Printed in The Netherlands. 0378-1909/81/061-059 $01.20.

reproductive strategy[1] in a single population has been modeled and found, under certain circumstances, to be evolutionarily stable (Gadgill 1972). Examples of obligate alternative strategies, where individuals practice only a single option, are rare in nature, perhaps because individuals practicing both reproductive options are at a competitive advantage over those which practice only a single option. All obligate reproductive strategies, however, need not result from balanced genetic polymorphisms. It is conceivable that an individual could be 'locked' into one developmental pathway or another entirely due to non-genetic factors.

This report focuses on a female mimicry reproductive pattern adopted by small male bluegill sunfish which has been previously shown to be obligate (Dominey 1980). Bluegill males in Michigan populations also mimic female behavior (Taylor 1979), and reports of multiple 'females' on nests in Colorado populations (Avilla 1975), and in a related species, *Lepomis megalotis* (Boyer & Vogele 1971) may also be accounts of female mimicry. Female mimicry may, in fact, be much more common in centrarchid fishes than indicated by the few reported instances.

The costs of female mimicry relative to typical male behavior are discussed, and two hypotheses are presented to explain the tolerance and 'courtship' of female mimics by the much larger, aggressive territorial males.

Materials and methods

Behavioral observations were made during the summer of 1976, 1978, and 1979 while snorkeling or scuba diving in Lake Cazenovia, N.Y; a 475 ha, mesotrophic, dimicric lake in upstate New York. The Lake Cazenovia fish community consists of 18 common species (6 centrarchids) and is typical of lakes in the region (Helfman 1979). Hundreds of hours of underwater observation were performed each summer on various aspects of bluegill reproductive biology, and 25 breeding colonies were monitored for spawning activity during 1978 and 1979. Dry, exposure suits permitted long periods of observation. Visibility and observation conditions were excellent and the fish were very robust to disturbance. Following all observations where female mimicry was suspected, individuals were captured with a hand net and stripped of gonadal products to verify their sex. Female mimics readily produced copious quantities of sperm upon slight pressure to the abdomen. The fertility of this sperm was tested in the laboratory, and the artificially fertilized offspring of female mimics were raised to freeswimming.

All data were recorded while underwater with pencil on plastic impregnated paper held in a clipboard. To determine the frequency of nest defense, 26 h of 10 min observation periods of active focal nests were taken. The importance of size in agonistic encounters was determined by observing several nests simultaneously during colony formation, and visually comparing the sizes of intruding and resident males during the 21 observed nest takeovers. The accuracy of such visual comparison was verified previously by capturing and measuring 20 pairs of nesting males after visual comparison; no errors in size ranking (by length) were made.

To determine the numbers of spawning neighbors for nonspawning and spawning nests, nest numbers were randomly selected; nests were permanently marked with numbered plastic disks. The condition of the nest (spawning or nonspawning) and the number and condition of adjacent nests were recorded. No nest was used twice. Following the initial selection, numbers were selected until a nest in a different condition (i.e. spawning if previous nest had been nonspawning) was selected and the process repeated until 20 nests in each condition had been censused.

Results

The bluegill mating system

Male bluegills in Lake Cazenovia occupied nesting territories in densely packed colonies of 15 to 500 active nests during the spawning period. Males competed aggressively for positions within these

[1] 'Strategy' is used to refer to a collection of adaptive responses accumulated over evolutionary time, and does not imply teleology (Thornhill 1979, Wilbur et al. 1974). Additionally 'tactics' or 'patterns' might better be used when individuals practice two or more reproductive options over the course of their lifetime, with 'strategy' reserved for the sum of all lifetime reproductive tactics or patterns.

colonies, and the ability to maintain residency on a nest was related to size. Males defended their nests against intruding adult males an average of 3 times per 10 min observation period. Nest defense most frequently consisted of swimming at the intruding male, however, occasionally contact was made, and in some cases prolonged fighting was observed. In data where male size was estimated, changes in nest residency due to male-male competition were recorded 21 times, of these 19 were instances of larger males displacing smaller males (Chi Square, $p<0.01$). Nearly all displacements involved vigorous, often prolonged fighting.

Males which mimicked female behavior ($\bar{x} = 126$ mm, s.d. $= 9$ mm, $N = 20$) were much smaller than nesting males ($\bar{x} = 190$ mm, s.d. $= 13$ mm, $N = 45$), and smaller males were never seen to compete aggressively for nests. Although the size distribution of female mimics was disjunct from that of nesting males, it did overlap with the lower end of the distribution for spawning females ($\bar{x} = 156$ mm, s.d. $= 19$ mm, $N = 45$). Because of the tendency for large males to displace small males, it is very unlikely that a male the size of a female mimic could maintain residency in an occupied portion of a bluegill colony.

Once males were established on nests, females arrived in large numbers and remained in the water column until spawning was initiated. Within a colony, spawning is highly synchronous, with the spawning period for an entire colony cycle occuring in a one to two day period. As (Hunter 1963) in the green sunfish, *Lepomis cyanellus*, bluegill females were attracted to areas of spawning activity, or 'spawning centers', and formed aggregations above the spawning nests. As a result of this female attraction to areas of spawning activity, nests adjacent to spawning nests had a higher probability of having spawning females present than did nests adjacent to nonspawning nests (Fig. 1). Males located away from these spawning centers frequently failed to receive eggs, and males which did not receive eggs, or received insufficient eggs to merit parental care (Dominey unpublished), abandoned their nests and traveled to other spawning colonies. At some colonies as few as 10% of those males occupying nests during the spawning period received sufficient eggs to merit parental care. An obvious result of the female preference for areas of spawning activity is that males providing parental care were found in more tightly clustered groups than would be the case if females indiscriminately selected nests. For females, this clustered spatial arrangement is highly adaptive because eggs placed in nests in areas with several actively guarding males are better protected from egg predators than are isolated nests (Dominey, manuscript).

To initiate spawning, a female leaves the water column aggregation and approaches a nesting male while displaying a coloration pattern (dark background, dark eyes, and pronounced vertical barring) which is known to reduce aggression in the nesting males of another sunfish (Colgan & Gross 1977). Once the female is in the nest, the male begins to swim in tight circles, turning, with the

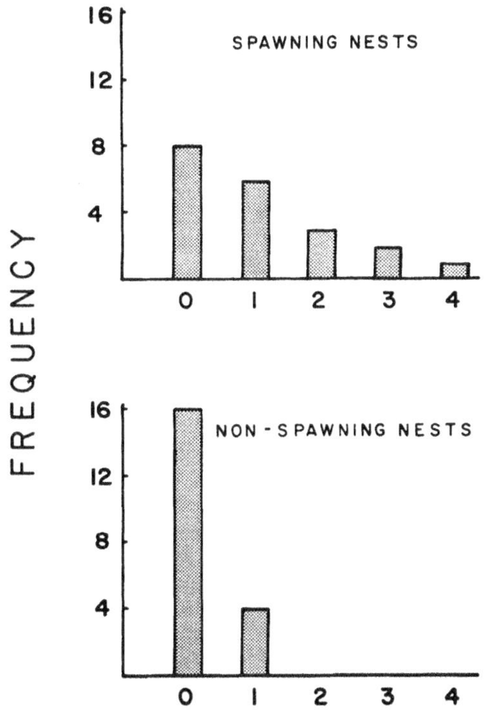

Fig. 1. Following sequential random selection of spawning and non-spawning nests, neighboring nests were censused for spawning activity. The frequency of neighboring nests having spawning females present is given. The data show an increased probability of spawning activity given that a neighbor is spawning. The two distributions differ (Mann Whitney, $p<0.01$), a result not due to a difference in the absolute number of neighbors present, which did not differ statistically ($p>0.5$).

female following. Every few seconds as the pair turnes, the female turnes on her side, presses her genital pore against that of the male, quivers, and releases eggs that the male fertilizes (Miller 1963).

The female mimicry reproductive pattern

Female mimics are small, sexually mature males which, like females, did not attempt to construct or defend nests, and were found in the water column aggregations. These small males were observed to leave the water column, adopt the dark, barred coloration, approach the aggressive territorial males without hesitating or retreating, and 'courtship turn' in the nests with resident males (Fig. 2). Such male-male pairs were observed to courtship turn for 10 min or more in the absence of a functional female. Female mimics are extremely similar to functional females in behavior and appearance. Female mimics, however, did not interrupt courtship turning with the lateral tilting and quivering characteristic of egg release. Also these small males tended to retain the dark coloration in the water column and did not have a 'gravid' appearance. Failure of an individual to show egg-release behavior during courtship turning was a reliable indicator of female mimicry, although occasionally functional females will courtship turn for short periods without showing egg-release behavior.

Although females mimics were observed to approach nesting males when no female was present, female mimics most often entered nests which already contained spawning pairs. Nesting males made little attempt to drive these intruders from their nests and behaved very much as if two functional females were present. Nesting males did seem to prefer to align themselves with functional females during courtship turning, and also seemed to direct more aggression at female mimics than at functional females; however, no quantitative data were taken. This difference is perhaps due to the gravid appearance and larger size of most functional females providing a better female stimulus to nesting males. Once female mimics were on the nest, the resulting trio, or occasionally larger group including several female mimics, would turn together in the nest, and spawn simultaneously. During egg release, female mimics were often located between the nesting male and female (see Dominey 1980). Since the sperm of female mimics was found to readily fertilize bluegill eggs in the laboratory, it is likely that during these spawning bouts female mimics were fertilizing eggs. Any eggs fertilized by female mimics are abandoned to the care of the nesting males.

Although female mimicry was seen at virtually every colony where spawning was observed, the frequency of occurrence was highly variable between colonies. On average, six percent of 'spawning nests' contained one or more female mimic.

Fig. 2. Nesting male and female mimic courtship turning on a nest in the absence of a functional female.

Discussion

Female mimicry should not be confused with the type of nest intrusions reported first for *Lepomis megalotis pelsaltes* in which males rush between spawning pairs just at the moment of egg deposition (Keenleyside 1972). In bluegills, males practicing this type of nest intrusion behavior are attacked by nesting males rather than courted or tolerated (Gross 1979), and as in *L. megalotis pesaltes*, any male, including neighboring nesting males, is capable of performing this behavioral pattern. Nest intrusion behavior, however, does appear to be a normal ontogenic interval of the reproductive strategy of female mimicry (Dominey unpublished).

As an obligate alternative reproductive strategy, female mimicry may be maintained in the population as a genetic polymorphism such as envisioned by Gadgill (1972). As such, the cost of female mimicry relative to nesting male behavior is of critical importance. The nesting male reproductive pattern is typical of highly sexually selected reproductive patterns. Nesting males compete aggressively for mating sites, display vigorously to attract mates, are sexually dichromatic, and show a higher variance in reproductive success (number of eggs in nests) than do the nesting males of related centrarchids (Carbine 1939). Large body size, sexual dichromatism, nest construction, fighting, and display must have associated costs. In addition nesting males provide 8–10 days of parental care during which time they repel potential nest predators, do not leave the nest to forage, and lose body weight (Dominey unpublished). The costs should be measured in terms of loss of future reproductive potential, and the cost of the nesting male reproductive pattern must be considerable.

In contrast, female mimics do not compete aggressively for nesting sites, display to attract females, show pronounced sexual dichromatism, maintain a large body size, or provide parental care. If female mimicry is a lower cost reproductive option, as seems likely, then female mimicry fulfills most of the requirements of Gadgill's model. The nex logical step would be to determine if the expression of this alternative reproductive pattern is inherited.

Regardless of whether female mimicry has a genetic basis, female mimics as individuals are in competition with nesting males, and if they are stealing fertilizations, why are they tolerated? This is a critical question in many cases of alternative reproductive patterns that can be considered 'parasitic' on the dominant male pattern. In the case of bluegills, guardian males may be deceived as to the sex of female mimics. This is generally consistent with what is known about mate recognition in other sunfishes, where mates are primarily recognized by their swimming movements (Noble 1934). However, there are other cues, e.g. the absence of egg-deposition movements, non-gravid appearance, or olfactory cues, by which nesting males could potentially recognize the sex of their spawning partners. All else being equal, selection should favor those nesting males who do not share their reproductive success with female mimics. Such selection may be absent because nesting males might reduce their fitness more by driving away female mimics during the courtship of functional females, than by sharing fertilizations. Female mimics can be persistent in reentering nests, and if nesting males interupt spawning frequently to drive away intruders, they risk the functional female leaving the nest. Thus a nesting male who shares part of his fertilizations might be as successful as a male who discriminates strongly against female mimics, and as a result loses potential fertilizations.

Another hypothesis is that nesting males may actually benefit from the presence of these smaller males, especially in cases where the nesting male has not yet attracted a female. Since females prefer to deposit their eggs in nests adjacent to spawning nests, any behavior that would cause a male's nest to appear to be a center of spawning activity should be favored – including turning in the nest with another male behaving as a female. For nesting males, then, the disadvantages of sharing incoming females with female mimics may be balanced with the advantages of producing a spawning center.

If the presence of female mimics does indeed benefit nesting males, then the situation might be analogous to that found in some lekking birds where territorial males tolerate the presence of satellites (Hogan-Warburg 1966), or display along with subordinates (Foster 1977). In both cases, territories which contain additional males are thought to be more attractive to females.

Neither hypothesis, deception or mutual gain, explains why female mimics should approach nesting males when females are not present in the nest. For a female mimic there is no obvious advantage to producing spawning centers since they can swim freely to areas of spawning activity. It may be, however, that there is considerable variation in the tolerance of nesting males for female mimics, and it might be advantageous to the female mimic to approach a tolerant nesting male in advance. Likewise, nesting males may be more tolerant of particular female mimics following 'courtship' behavior with them.

Acknowledgements

This research was supported by a Doctoral Dissertation Improvement Grant from the National Science Foundation, and by grants from the Ithaca and

National chapters of Sigma Xi, the George D. Harris Foundation, and the Theodore Roosevelt Memorial Fund of the American Museum of Natural History. Use of the Cazenovia College Lake Station was obtained by D. Roy, director, and L. Olney granted permission to use SUNY equipment present at the station. G. S. Helfman and E. B. Brothers introduced me to underwater field techniques. D. Knaack and P. W. Colgan gave comments on the manuscript. A. Donjacour, A. Horn, S. Letovsky, C. Mercaldi, A. Plassman, C. Van Buren and J. Weiskotten served as field assistants.

References cited

Avilla, V.L. 1975. A field study of nesting behavior of male bluegill sunfish (*Lepomis macrochirus* Rafinesque). Amer. Midl. Natur. 96: 195–206.

Boyer, R.L. & L.E. Vogele, 1971. Longear sunfish behavior in two Ozark reservoirs. pp. 13–25. *In*: Hall. G.E. (ed.) Reservoir Fisheries and Limnology, Spec. Publ. No. 8, Amer. Fish. Soc., Wash., D.C.

Carbine, W.F. 1939. Observations on the spawning habits of centrarchid fishes in Deep Lake, Oakland County, Michigan. Trans. N. Amer. Wildl. Conf. 4: 275–287.

Colgan, P.W. & M.R. Gross. 1977. Dynamics of aggression in male pumpkinseed sunfish (*Lepomis gibbosus*) over the reproductive phase. Z. Tierpsychol. 43: 139–146.

Dominey, W.J. 1980. Female mimicry in bluegill sunfish – A genetic polymorphism? Nature 284: 546–548.

Foster, M.S. 1977. Odd couples in manakins: A study of social organization and cooperative breeding in *Chiroxiphia linearis*. Amer. Natur. 111: 845–853.

Gadgil, M. 1972. Male dimorphism as a consequence of sexual selection. Amer. Natur. 106: 574–580.

Gross, M.R. 1979. Cuckoldry in sunfishes (*Lepomis:* Centrarchidae). Can. J. Zool. 57: 1507–1509.

Helfman, G.S. 1979. Twilight activities of yellow perch, *Perca flavescens*. J. Fish. Res. Board Can. 36: 173–179.

Hogan-Warburg, A.J. 1966. Social behavior of the ruff *Philomachus pugnax* (*L.*). Ardea 54: 109–229.

Hunter, J.R. 1963. The reproductive behavior of the green sunfish, *Lepomis cyanellus*. Zoologica 48: 13–24.

Keenleyside, M.H.A. 1972. Intraspecific intrusions into nests of spawning longear sunfish (Pisces: Centrarchidae). Copeia 1972: 272–278.

Miller, H.C. 1963. The behavior of the pumpkinseed sunfish, *Lepomis gibbosus* (Linnaeus), with notes on the behavior of other species of *Lepomis* and the pygmy sunfish, *Elassoma evergladei*. Behaviour 22: 88–151.

Noble, G.K. 1934. Sex recognition in the sunfish. *Eupomotis gibbosus* (Linne). Copeia 1934: 151–154.

Taylor, J.N. 1979. Behavioral components of reproductive success in male sunfishes of the genus *Lepomis*. Ph.D. thesis, University of Michigan, Ann Arbor. 254 pp.

Thornhill, R. 1979. Adaptive female-mimicking behavior in a scorpionfly. Science 205: 412–414.

Wilbur, H.M, D.W. Tinkle & J.P. Collins. 1974. Environmental certainty, trophic level, and resource availability in life history evolution. Amer. Natur. 108: 805–817.

Env. Biol. Fish. Vol. 6, No. 1, pp. 65-85, 1981

Patterns of parental investment, dispersal and size among coral-reef fishes

George W. Barlow

Department of Zoology & Museum of Vertebrate Zoology, University of California, Berkeley, California 94720, U.S.A.

Keywords:
Eggs, Fecundity, Food, Freshwater fishes, Larvae, Open coast, Ovoviviparity, Patchiness, Plankton, Predation, Viviparity

Synopsis

Coral-reef fishes have been selected to produce propagules for dispersal because they live in a patchy environment, and the adults cannot migrate between patches. For large species (> 100 mm SL) and widely separated patches, numerous propagules are needed, often with specialized pelagic intervals. Individuals of small species are confined to portions of the reef. They are unable to produce enough eggs for effective long-range dispersal, and so they keep their vulnerable eggs and young out of the plankton until they are well developed enough to seek out and settle onto the appropriate habitat before dispersing.

Guarded demersal eggs, requiring a greater individual investment by the small short-lived species, further reduces their individual fecundity. These costs to fecundity, and the reduction in vagile young appear to account for the lack of postzygotic investment in larger longlived species.

All coral-reef fishes are selected to disperse, usually with a young planktonic propagule interval. Both large and small species produce a mixed outcome, with some propagules returned to, or retained at the home reef, while others disperse more widely. The smaller the species the greater the proportion of propagules retained.

Introduction

In comparison to freshwater species, coral-reef fishes seem to favor broadcasting large numbers of planktonic eggs. Why this is so, however, and whether it is even a reasonable generalization, have received little critical attention until recently.

In his review from the literature and his own experiences, Johannes (1978) promoted the belief that coral-reef fishes characteristically release immense numbers of small eggs that are borne away from the reefs on efferent currents. To account for what he termed this *r* strategy of reproduction he considered three hypotheses: (1) the *food hypothesis* – feeding for the young is better away from the reef; (2) the *anti-predator hypothesis* – predation on the reef is too intense for the survival there of eggs, embryos and larvae; and (3) the *dispersal hypothesis* – the young stages are propagules whose primary function is dispersal. Johannes concluded that avoidance of predation on eggs, embryos and larvae explains the pattern of reproductive behavior shown by coral-reef fishes.

Received 28.4.1980 Accepted 5.9.1980

I argue rather that dispersal is the primary

© Dr. W. Junk b.v. Publishers, The Hague. Printed in The Netherlands. 0378-1909/81/061-065 $04.20.

selective force molding the reproductive behavior of coral-reef fishes. Predation on eggs and larvae is indeed a powerful selective force, but it does not account for the predominance of planktonic propagules. Moreover, Johannes was overly concerned with species that are relatively large and conspicuous, a natural consequence of a literature that focusses on such species, and of his interactions with native fishermen. Sale (1978a) distinguished between large and conspicuous ('visible') species and small inconspicuous ('invisible') ones; but since many small species are highly visible, and some large ones not, that terminology will not be followed.

My approach relates the different patterns of investment per offspring among coral-reef fishes to size of fish. From this emerges a general view that large species are indeed profligate broadcasters but that small species cannot play the same game because of their limited fecundity and lesser mobility. Nonetheless, the need for larval dispersal, even if limited, is still held to be a central factor.

I will compare large and small species as two discontinuous categories although a spectrum exists. I take as a reasonable point of separation a standard length (abbreviated SL) of 100 mm. Length is not the best indicator of size, but weights are too seldom published. Thus, a rotund boxfish or sculpin 100 mm SL might best be considered as large, but a slender pipefish 200 mm SL as small. Also, by fecundity I mean number of eggs, not the proportion of the fish taken up by gonads, the meaning Williams (1966) employed. Postzygotic investment is any allocation of energy to the well-being of the zygotes, such as care of demersal eggs, mouthbrooding, or retention of fertilized eggs.

Reproductive behavior of coral-reef fishes

Egg broadcasters

In this pattern (reviewed by Johannes 1978), a female and one or more males come together briefly and release sperm and planktonic eggs. Representative families include the Serranidae, Mullidae, Pomacanthidae, Chaetodontidae, Acanthuridae, Labridae and Scaridae. Most of the mobile species make brief spawning migrations to the edge of the reef, often around a pass, or to the downcurrent side of a seaward extension of the reef. The exact place may vary with the current (Warner et al. 1975). Even small wrasses such as *Thalassoma bifasciatum* swim as far as 500 m to spawn (Robertson & Hoffman 1977). The timing is important, spawning being synchronized with an outflow through the pass or with the off-flow from the reef that commences at high tide. The female and the male(s) usually spawn at the apex of a rush upward from the bottom; then they dash back down to the reef. This complex ensemble of behavior seems an adaptation to releasing planktonic eggs up in the water column where they will be transported away from the reef (Randall 1961, Randall & Randall 1963, Barlow 1975a, Choat & Robertson 1975, Warner et al. 1975, Colin 1978, Colin & Clavijo 1978, Johannes 1978, Lobel 1978). Variations on this theme are evident, and probably more will be observed (Barlow 1975a, b, Robertson & Hoffman 1977, Colin 1978, Lobel 1978, Moyer & Nakazono 1978a, Fischer 1979).

Since mobility decreases with size, very small (i.e. < 50 mm SL) species generally spawn where they reside. For example, basslets of the genus *Liopropoma* (Randall 1968) perform brief, short upward dashes in pairs from their home sites, apparently to spawn (Eric Fissher, personal communication). They are probably also territorial, as are nearly all such small bottom-dwelling species. Most very small species that live in intimate association with the reef, such as gobioids and blennioids, have a morphology not suited to migrating, and would be vulnerable to predators if they swam up to spawn. Parental care of eggs is prevalent in fishes tied to the substrate, though it occurs in other groups as well.

Postzygotic investment

Parental investment after fertilization seems limited to small fishes with demersal eggs. Although fishes that broadcast their eggs sometimes briefly chase fishes that approach to eat their floating eggs (Robertson & Hoffman 1977, Fischer, personal communication) such defense is too sporadic and brief to be comparable to the caretaking shown by species without demersal eggs.

The most common form of parental care is that

directed to a hidden clutch of adhesive eggs. Caretaking ceases at hatching, when the young ascend into the plankton. The male is the sole caretaker when only one parent is involved. This behavior characterizes a number of families that consist mostly of small bottom-dwelling species: the Gobiidae, Eleotridae, Gobiesocidae, Plesiopidae, Blenniidae, Clinidae, Chaenopsidae and Trypterygiidae are examples. It has been suggested on the basis of adhesive eggs that one surgeonfish in Japan, *Teuthis fuscescens*, engages in parental care (Breder & Rosen 1966), but that is not a necessary conclusion.

The Pomacentridae (damselfishes and anemonefishes) also belong here (reviewed by Reese 1964, Fishelson et al. 1974). The eggs of some are out in the open, as in *Chromis caeruleus* which spawns on tufts of algae. Also, they are active swimmers, and none of the species rests on the bottom, though they refuge in the reef. Algal-feeding species stay close to the bottom, but the larger planktivorous ones swim well up from the reef. In either case, males display to females and chase potential egg predators making them highly visible (personal observation).

Monogamous pairs form in the caretaking anemonefishes of the genus *Amphiprion* (Allen 1972, Fricke 1976, Ross 1978), and with a few smaller associates have as their territory a large anemone, where they spawn next to its foot. When the embryos hatch, parental care ceases and the young become planktonic. Monogamous pairs have also been reported in some small gobies (Schultz 1948, Lassig 1976, 1977).

Parental care extending beyond hatching, to include larvae, is exceedingly rare among coral-reef fishes. So far, only the case of the small Australian damselfish, *Acanthochromis polyacanthus* is known (Robertson 1973). The male and female care for about 100 (30–190, Allen 1975) eggs, which are most likely large because they are so few. The free-swimming young are guarded for a 'few' months, during which they grow to 28 to 40 mm in length. The young appear initially to feed from the mucus on the surface of the parents. When parental care ceases, the young take up residence on the reef, omitting the planktonic interval.

Alternative avenues to postzygotic investment, other than guarding embryos or larvae, exist. Many species bear their embryos or young (Balon 1975a), whether within or attached to the parent. Apparently all the cardinalfishes (Apogonidae) are mouthbrooders, predominantly with male care though mouthbrooding females have been reported (Breder & Rosen 1966, Oppenheimer 1970, Chave 1971, Smith et al. 1971, Hobson & Chess 1978). The jawfishes (Opistognathidae) are also paternal mouthbrooders; these small fishes excavate intricate burrows in the sand where they take refuge (Böhlke & Chaplin 1957, Böhlke & Thomas 1961, Leong 1961, Colin 1972, Colin & Arneson 1978, Kerstich 1979).

The Syngnathidae (pipefishes and seahorses) are small reef species in which the males bear the embryos in a ventral brood pouch. In the closely related monotypic family, the Solenostomidae, the female retains the fertilized eggs in a pouch formed by her pelvic fins (Fishelson 1966). After hatching, the young are released into the plankton, though the young of some temperate-zone pipefishes and seahorses probably lack the pelagic interval (e.g. Russell 1976). On the other hand, Williams (1960) found that young of a North Atlantic pipefish are especially effective dispersers.

Among the ophidioids there are viviparous and ovoviviparous species; apparently the larger species give birth to young that become planktonic (e.g. Wourms & Bayne 1973). Some of the clinids bear their fertilized eggs internally (Breder & Rosen 1966, Penrith 1969). In at least some cardinalfishes the eggs are fertilized within the female where they are retained for a few hours before she gives them to the mouthbrooding male (Chave 1971). There are a few old reports (cited in Breder & Rosen 1966) of internal fertilization among some tropical gobies, but they are probably erroneous.

Fecundity of coral-reef fishes

Coral-reef fishes, regardless of pattern of parental investment, are characterized by great fecundity and a virtually continuous breeding season (Randall 1961, Munroe et al. 1973, Warner et al. 1975, Luckhurst & Luckhurst 1977, Johannes 1978, Sale 1978). In general eggs are small, which increases the number of eggs per clutch, even in species that invest postzygotically. For example, the smallest apogonids (about 35 mm LS) have around 1,500 eggs (Chave 1971, Carney 1976). By comparison, a freshwater mouthbrooding cichlid of the genus *Labeotropheus* of comparable size carries 30 – 50 eggs, each nearly five millimeters in diameter (Balon 1977).

Selection for high fecundity probably also explains why in some monogamous species of coral-reef fishes the female is larger than the male, which contrasts sharply with comparable freshwater fishes. In the anemonefishes the female is the largest individual of a reproducing unit (Allen 1972, Fricke 1976, Moyer & Nakazono 1978b, Ross 1978) and the male *Amphiprion* does most of the caretaking, which allows the female to continue feeding and thereby develop more eggs (Allen 1972; see also Qasim 1956, Loiselle & Barlow 1978). In the peculiar little *Solenostomus cyanopterus*, however, it is the female who carries the embryos, and she is huge compared to her mate (Fishelson 1966). The female is much the larger sex in monogamous pairs of the surgeonfish *Acanthurus leucosternon;* a pair is better able to defend a feeding territory, and paired females have more fat reserves than do solitary ones (Robertson et al. 1979).

The general pattern that emerges for coral-reef fishes, then, is one of relatively low investment per offspring. The same conclusion holds, but with less force, for those species that invest postzygotically.

Size distribution of coral-reef fishes and postzygotic investment

The foregoing has indicated that egg-broadcasting species are large while those that invest postzygotically are small. If small species are inclined to invest postzygotically, then one might expect that to be a common pattern simply because teleost species are predominantly small (Marshall 1971; but see Lindsey 1966). Molles (1978), for instance, reported that 72% of the reef inhabitants near Guaymas, Gulf of California, were mostly small species. Burgess & Gilbert (1978) concluded that Caribbean reef species are primarily small, though they did not mention precautions taken to avoid biased sampling.

I tested two related hypotheses: (1) most coral-reef species are small, (2) postzygotic investment is about equally distributed among species of different sizes. Most reports of collections are of limited use because of bias inherent in sampling from a community which has at one extreme highly evasive species and at the other secretive forms that withdraw into the interstices of the reef.

Clark et al. (1962) presented an exceptional analysis of the abundance of coral-reef fishes and their sizes, gathered both by visual census and by collecting specimens with an ichthyocide but cautioned that their study site in the Red Sea was relatively sparsely inhabited by fishes. The majority of species reported, 68%, were 100 mm SL or less (greatest length listed for each species) (Fig. 1).

Randall (1963) captured all the fishes from two similar shore reefs on St. John in the Virgin Islands by enclosing them with large nets, applying an ichthyocide, and colllecting all day with a team of helpers. Slightly more than half the species were 100

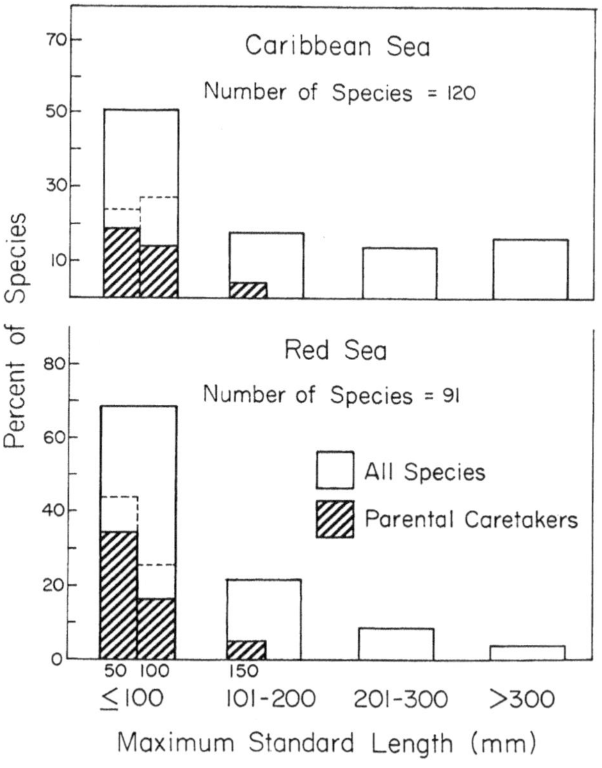

Fig. 1. The open columns show the percent of all the fish species of that size class on a coral reef. The columns with diagonal lines represent the percent, of all the species, that has parental care. Within the size class ≤ 100 mm SL, the species have been further divided into two subclasses, but the enclosing dashed line indicates the percent of species ≤ 100 mm SL. The percent of caretaking species in the class 100–150 mm SL is also shown (after data in Clark et al. 1962 and Randall 1963).

mm SL or less (Fig. 1). Aside from the greater abundance of large fishes such as groupers and the refuging pomadasyids and lutjanids, the distributions of fish lengths are remarkably similar to those of Clark et al. (1962). It appears that most coral-reef species are indeed small.

I then assigned the various families in these reports (Clark et al. 1962, Randall 1963) to two categories, 'parental care' or 'egg broadcaster'. Species assigned to 'parental care' are ones in which the species in that family so far reported show such care. 'Egg broadcaster' was similarly designated, except that when nothing of the family's behavior was known the species was put into this category. That biases the data in favor of broadcasters.

The results (Fig. 1) reveal that parental care is confined almost entirely to small species (≤ 100 mm SL). Moreover, most of the small species, 74% of them in the Red Sea collection and 67% in the Caribbean, engage in such caretaking. Of the species 50 mm SL or less, 77% and 79% respectively, are caretakers; of those between 50 mm and 100 mm SL, 68% and 53%, respectively, are caretakers; only one species longer than 100 mm SL had parental care in the Red Sea study, and only five in the Caribbean – all were less than 150 mm SL.

Thus postzygotic investment is not uniformly distributed across species, but is confined mostly to small ones. And the data suggest that the smaller the species the greater is the probability that it has evolved some form of caretaking.

The small species lacking parental care stand out as exceptional. Some of the small species were merely juveniles of larger forms tallied as small species to avoid bias where I was uncertain of allocation. Among the Red Sea fishes, two of the small species were trunkfishes (Ostraciontidae) and two were scorpionfishes (Scorpaenidae); all four are rotund species that might contain more eggs than their lengths would suggest (see Vitt & Congdon 1978). Both collections contained some small plectognaths, e.g. filefishes, whose reproductive behavior is simply unknown. The foregoing species are thought to be gonochorists (separate male and female sexes); further studies, however, may prove otherwise.

The remaining small fishes appear to be protogynous hermaphrodites. Among the small species in the study by Clark et al. (1962), 74% engage in parental care and 15% are broadcasting hermaphrodites, for a total of 89%. The remaining 11% are either broadcasting gonochorists or are unknown. The proportion of small species that are broadcasting gonochorists will probably sink below 11% as they become better studied, making such species particularly worthy of attention. A similar trend is apparent in Randall's (1963) report; most of the small species that do not invest postzygotically are protogynous wrasses or hermaphroditic serranids and allied families.

Size and vagility

Vagility refers to the capacity to move about, whether by active or passive transport. Some organisms may have limited locomotory ability but still be vagile, such as individuals rafting in algae, and planktonic eggs or young. In adults however, vagility usually corresponds to mobility, and hence size (Beamish 1978).

Most coral-reef species, even those feeding on plankton, seek shelter in the reef and are therefore not highly mobile as adults. Some have modestly large home ranges, however, such as many chaetodontids (Reese 1975, Ehrlich et al. 1977), and sugeonfishes (Barlow 1974). But the more a species lives and feeds within the reef itself, the smaller and less mobile it tends to be. Some groupers, snappers, grunts and similar species refuge during the day in or over the reef, but they move considerable distances to feed at night, even well off the reef into sandy areas (Hobson 1965, Ogden & Ehrlich 1977, Hobson & Chess 1978).

Why broadcast pelagic eggs?

Broadcasting pelagic eggs does not typify all reef species. Yet species that broadcast comprise up to half those of the reef. Why do they broadcast their eggs? The discussion that follows refers primarily to large species.

Food hypothesis

This hypothesis assumes that the open ocean is a more favorable feeding place for the young than is the reef. This could be either because there is too much competition on the reef, or because too little edible plankton is produced there.

The open sea is poor in plankton (Johannes 1978), and although it can be locally abundant in large though weakly predictable patches (Hunter 1976), there is no evidence that fish larvae can seek out such patches. Downcurrent from islands and atolls there are sometimes gyres where plankton is concentrated (references in Johannes 1978). If a larva were to remain within the lagoon of an atoll, however, it would find plankton relatively more abundant (references in Johannes 1978). It would thus seem best to put eggs into the lagoon, as a few species do (e.g. Helfrich & Allen 1975), rather than into currents flowing away from the reef on the chance the larvae will land in a favorable gyre.

It may well be that the most favorable situation for larval feeding, irrespective of size of fish, involves adopting a pelagic mode of life, but not necessarily one that entails moving out into the ocean. The food hypothesis can be rejected as an explanation for the offshore movement of the eggs, embryos and larvae of the large species of coral-reef fishes.

Anti-predation hypothesis

This hypothesis assumes that the coral-reef community feeds heavily on fish eggs, embryos and/or larvae. Mass spawning would swamp local egg predators, as might synchronous hatching of embryos by damselfishes (Keenleyside 1972, Holzberg 1973), and nocturnal hatching of guarded embryos (Allen 1972, Fricke 1976, Moyer & Bell 1976, Ross 1978), and nocturnal settling of larvae (Hobson & Chess 1978, Johannes 1978) can be reasonably viewed as anti-predation tactics.

Sessile filter feeders can remove up to 50% of the plankton in a few hundred meters (Johannes 1978), including the larvae of damselfishes (Fricke 1976). Planktivorous fishes form clouds over the reef where the currents are strong and may even congregate downcurrent from spawning parrotfish and move with the cloud of eggs (Colin 1978). Fish eggs were ranked the second or third most abundant items found by Hobson & Chess (1978; see also Meyer 1977) in the guts of six of nine planktivorous pomacentrids, though fish larvae were noted in only three species. Since the piscine planktivores that eat eggs are themselves small species and need to stay near the reef for refuge (Hobson & Chess 1978), once the eggs are away from the immediate vicinity of the reef they are presumably less subject to predation. Thus directly offshore, in the sea or in a lagoon, might be a relatively safe haven.

Johannes (1978) and Lobel (1978) interpreted this evidence as favoring the anti-predator hypothesis: The eggs are said to escape the reef planktivores to grow in the sea until they can return as juveniles. The evidence does suggest that this early predation is a strong selective force, but it does not explain the general pattern of parental investment among coral-reef fishes. Nor does it reveal whether levels of such predation in the open sea are actually lower than over the reef. The rate at which eggs, embryos and larvae perish in the open sea is startlingly high, up to 29% per day for eggs, and ten percent per day for larvae (Jones & Hall 1974, Ware 1975), and since the mortality rate is thought to be positively correlated with temperature (Ware 1975), it is probably higher in the tropics.

The chief difficulty with the anti-predator hypothesis, however, is that it fails to consider alternative adaptations to avoid predation on the young. For instance, why not lay a few large eggs and guard them and the ensuing young, as is done by at least one damselfish, rather than produce millions of eggs, over most of the year. Why not become viviparous? Viviparity has evolved repeatedly in freshwater fishes (Lowe-McConnell 1975), and to a high level in surfperches (Embiotocidae) of the North Pacific (Shaw 1971, Bryant 1977). There must be some reason other than simply minimizing predation to explain why the large coral-reef fishes release so many eggs into the open sea.

Dispersal hypothesis

The distribution of fish species across the tropical Pacific provides overwhelming evidence of dispersal. For example, a faunal treatment of the fishes of southern Japan serves well in identifying the bulk of large species in the Marshall Islands. The distribution pattern favors the hypothesis that large coral-reef fishes are actively dispersing their young.

One might argue, nonetheless, that dispersal is merely a byproduct of some other adaptation.

Johannes (1978) rejected the dispersal hypothesis because certain findings suggested that reef fishes time their spawning to maximize the probability that their young will be brought back to their source. The central evidence, from several different geographical regions, is that peaks of spawning occur when major currents are the weakest, minimizing passive transport away from the place of origin, with the possibility that eggs drift into major gyres downcurrent from islands, eventually to return the larvae to their source (Jones 1968, Sale 1970).

The best studied area in this respect is Hawaii, which has two annual peaks of spawning coinciding with the two periods of weakest currents and the presence of major gyres. One genetic study indicates a degree of endemism between the Hawaiian Islands, which suggests local recruitment (Smith 1969). On the other hand, species introduced to one island have spread to others (Randall & Kanayama 1972).

The foregoing does not necessarily argue against dispersal. Rather, the findings about seasonality and the suggestion about the role of gyres (there are no data to assess their importance) may simply reveal some of the features of the pattern of dispersal consistent with bet hedging (Stearns 1976), a strategy of mixed outcomes (Gadgil 1971).

I am not suggesting that a genetic polymorphism is responsible for a mixture of outcomes. Rather, two factors could provide varying proportions of widely and locally dispersed propagules. By spawning most when the currents are weakest and a gyre is present, chances are that a larger proportion of the propagules will be returned; but spawning also occurs throughout the year.

The second factor is the nature of the eggs and young. The young emerging from planktonic eggs are small; but since the pelagic interval is prolonged they increase in size before taking up life on the reef. Specialized pelagic stages can grow to as much as 30 mm SL (Breder 1949), for example the acronurus of the surgeonfishes, the tholichthys of the butterflyfishes, and the leptocephalus of eels. The longer in time (the further) an egg or larva spends traveling the lower its probability of surviving (Thorson 1946, Gadgil 1971). For successful longrange dispersal an individual must therefore produce great numbers of propagules. Those fishes that have successfully crossed the so-called East Pacific Barrier, going from the Indo-West Pacific to Panama, all have specialized larvae (Rosenblatt et al. 1972). Perhaps equally important, they are all large fecund species (see also Rosenblatt 1963).

Is there an optimum size of egg for dispersal?

If successful dispersal depended only on the number of propagules, then a fish ought to minimize egg size to maximize numbers. If a lower limit to egg size exists, it might have been reached by the broadcasters. There might be a minimum size at which the brain and sense organs can function (Haldane 1928), or the viscosity might be too great for tiny larvae to swim.

The smallest diameters recorded for marine fish eggs are about 0.4 mm (Hiemstra 1964, Breder & Rosen 1966, Marshall 1971, Russell 1976). Some apogonids, however, have yet smaller eggs, e.g. a mean diameter of 0.24 mm (Charney 1976), and some gobies have comparably small eggs (Breder & Rosen 1966, Marshall 1971). A number of freshwater species (e.g. the characins) also have such small eggs. Most planktonic fish eggs, however, are larger than those just mentioned, but there may be an over-riding effect of temperature (as is the case in Fig. 2). Within related groups, egg size is larger at higher latitudes (Marshall 1953, Russell 1976), and generally inversely correlated with surface water temperature (Williams 1967, Ware 1975, 1977).

Ware's (1975) model for optimum size of pelagic fish eggs (0.5 – 0.6 mm) indicates a trade-off, with increasing temperature, between the smaller offsprings' increased rate of mortality as opposed to the greater number of eggs made possible by their smaller size, and to their faster development. Size of pelagic eggs does vary between tropical marine fishes living under the same temperature regimen, so other selective forces must be operating.

There is abundant documentation that fecundity is proportional to female size within a given species (Svärdson 1949, Nikolsky 1963, Bagenal 1967, Weatherly 1972). There is sometimes a less significant tendency for eggs to increase in size with age of female (Pope et al. 1961, Bagenal 1967, Blaxter 1969, Ware 1977), but this does not alter the basic relationship. Since growth continues in adults in most species females become increasingly fecund with age (Williams 1966). This is particularly important to the larger species, since larger species of fishes tend to live longer (Rosenblatt 1963, Williams 1966).

Since individuals of each species should spawn eggs close to the optimum size, the number of eggs produced is then constrained by the size of the female. Thus for broadcasting numerous eggs, females should be greater than a certain minimum size.

Conclusions about dispersal

The foregoing evidence suggests dispersal as the primary selective agent causing large species to broadcast small eggs into offshore currents, with adaptations to minimize predation on those propagules near the reef. But since other means of minimizing predation on young have not evolved beyond isolated examples, and since those means are more common in other fish communities, I presume that the anti-predation adaptations are secondary. The anti-predation hypothesis has the further difficulty of being unable to explain why open-coast species situate their eggs and larvae inshore; a fuller understanding of the problem of dispersal in different communities can account for this difference (see below).

This moves the argument from why large coral-reef species broadcast eggs to why should they disperse, particularly by means of eggs and larvae? To respond to this question I need to discuss the nature of the coral-reef environment, and to compare the problems of dispersal there to those of coastal and freshwater fishes. In addition, since there are still differences of opinion among ichthyologists as to whether postzygotic investment is really more frequent, more diverse, and more highly developed among freshwater fishes, I will briefly review that issue.

The coral-reef environment

Atolls, fringing insular reefs, disjunct headlands, and disrupted coastal reefs, even if rocky rather than coral, all present fish species with the same type of problem, with some differences, but I will consider only coral reefs in detail. At their outer edges, coral reefs give way to the open ocean where there is no place to take refuge, plankton is sparse and food is scarce, and species living there are highly adapted to the open environment. Coral reefs consist of patches within patches. In the central Pacific, for example, there is a vast network of disjunct islands separated by open ocean. Similarly, coastal coral reefs are disrupted by channels and lagoons (e.g. Clark & Gohar 1953). At another level, a given reef is dissected by passes, sandy areas, or rubble, or is broken up into different types of corals.

There is disagreement as to whether a coral reef is a stable or an unstable environment, depending on whether stability is discussed with reference to competitors (Sale 1977, 1978b), to small patches, to an entire atoll, or to different spans of time. When large storms sweep coral reefs their impact can be dramatic (Robins 1957, Springer & McErlean 1962, Randall 1974, Bortone 1976). From personal observations at Enewetak Atoll, in 1972, I can attest to the reshaping of the reef by just the edge of a typhoon: New flow areas were cut and old ones filled with sand, some parts of the reef were torn away and thrown into the lagoon, forming foci of new patch reefs; heavy wave damage occurred inside of the lagoon as deep as 35 m; damaged fishes, particularly of the species that live close to the reef, were common.

Irrespective of storms, coral reefs reflect a continual conflict between growth and degradation (Goreau & Hartman 1963, Randall 1974, Adey 1978). Given the time scale within which evolution works, and the age of coral-reef fish communities, coral reefs are best regarded as undergoing continual change.

The large-scale patchiness of reefs precludes movement between reef systems of the adults of all species that are closely adapted to the reef, such as parrotfishes, wrasses, surgeonfishes, butterflyfishes and pomacentrids. Their only means of dispersal is pelagic propagules. There is strong selection, therefore, to produce vast numbers of eggs, to release them so as to be borne away, and to perfect specialized pelagic young, all to maximize the probability of settling within a different reef system.

I have no adequate answer to explain why dispersal is itself so important to fitness, though I will return to this issue in closing. It is uncomfortable to argue that instability within the reef system is the driving force because I am treating large species that have sufficient mobility to change residence within a reef system (but instability seems to be of obvious importance to small species). For the moment, the need to disperse is inferred from the finding that the large fishes are indeed widely distributed.

The open coast and its fishes

In contrast to coral-reef systems, large portions of marine coasts consist of vast stretches of undifferentiated sandy or muddy bottom; headlands and reefs are only seldom obstacles. Shore fishes that inhabit the open coast are adapted to life in an environment that lacks ready cover in one of two ways. Benthic species are adept at matching the background and are depressed. Often they are able to submerge, totally or partially, into the sand or mud without regard to a particular home site (e.g. flatfishes, weavers, batracoids, platycephalids, dactylocopids and ogcocephalids). There is also a fauna of burrow-dwelling fishes such as the garden and cusk eels, opisthognathids, branchiostegids, and gobioids. The other adaptation is to form schools that move over the bottom. These are strong swimmers, protectively colored by stripes, silvery hues, and schooling marks. They include the ariid catfishes, some of the mullids, sciaenids, haemulids, pomadasyids, polynemids, lethrinids, gerrids and sparids.

Johannes (1978) pointed out that many fishes in tropical marine environments move inshore to spawn. He mentioned the Gerridae, Sparidae, Engraulidae, Pomadasyidae and Sciaenidae, all of which are open-coast forms; many of these spawn in estuaries. The larvae of sciaenids and pomadasyids are seldom taken offshore, apparently settling out quickly and schooling over the bottom (Hildebrand & Cable 1934, Rosenblatt & Walker 1963, Robins 1971, Powles 1978).

Small fishes often take shelter in clam or shrimp burrows, or under shells. Typically, these are gobioid species or similar forms in which the male cares for eggs that hatch into pelagic larvae. The ariid catfishes, a secondarily marine family, are mouthbrooders (Gunter 1947, Breder & Rosen 1966) and their young are released as juveniles, without a pelagic interval. The opisthognathids are also mouthbrooders. The Dactyloscopidae carry their eggs behind their pectoral fins (Böhlke & Chaplin 1968, personal observations). The batrachoids (toadfishes) are sedentary large species on mud or sand bottoms (Gill 1907, Gudger 1910) in which the male guards the large eggs (ca. 5 mm diameter in *Opsanus tau*). The young remain benthic, never entering the plankton. In the Carribean Sea, males of *Amphichthys cryptocentrus* guard huge eggs 10 mm in diameter, and continue to protect the young in a depression in the sand (Hoffman & Robertson, in preparation); no information is yet available as to whether the young have a pelagic dispersal phase, though that seems unlikely.

The available evidence suggests a tendency for fishes of the open coasts to have less dispersive young than those of the coral reefs. Some, such as the batrachoids and ariids, and perhaps the ophidioid *Ogilbia cayorum* (Longley & Hildebrand 1941, Suarez 1975) and its relative *Gunterichthys longipenis* (Dawson 1971), omit the planktonic interval altogether.

Tropical freshwaters and their fishes

At the risk of oversimplification, I assert that freshwater fishes are primarily adapted to life in flowing water, lacustrine species being derived forms. The general pattern is for adults to move upstream to spawn and after the (usually demersal) embryos hatch for the young to return downstream, assisted by the current. This pattern of dispersal differs radically from that of coral-reef fishes in that both adults and young participate. That removes a major constraint on patterns of postzygotic investment. Producing large offspring does not preclude dispersal.

I will compare the occurrence of postzygotic

investment across families of tropical freshwater fishes. Ovoviviparity and viviparity are taken up further on. I have arbitrarily taken the family as assigned by the authority cited, and lumped all reef species and all freshwater species. Some allocations to categories of investment are based on fragmentary information; where no information is available the family is omitted.

This crude tabulation (Table 1) suggests that at least some parental care occurs in 54% of freshwater families, but in only about 20% of reef fishes. If one considers the degree to which parental care has evolved, the difference becomes even more apparent.

Among tropical freshwater teleosts, care of young after the start of exogenous feeding occurs in many groups. Within the Cichlidae, joint parental care of free-swimming young, lasting several weeks, is found in hundreds of species in South America, Africa and Asia. Joint parental care also occurs in *Heterotis niloticus* (Blache 1964) and *Arapaima gigas* (Lüling 1964). The male *Gymnarchus* guards the free-swimming young until they reach a length of about 75 mm (Breder & Rosen 1966). Many species of cichlids (e.g. *Apistogramma* and *Sarotherodon*) have prolonged parental care of the juveniles during which the female is the caretaker. Most of the species covered in this paragraph are small, though several are relatively large (e.g. *Boulongerochromis microlepis*, *Cichla ocellaris* and *Gymnarchus*). One, *Arapaima*, is huge, reaching lengths of two to three meters.

Mouthbrooding appears in at least five freshwater families (Oppenheimer 1970), and has been independently evolved in different genera within the same family (e.g. *Aequidens*, *Geophagus*, *Haplochromis* and *Sarotherodon* in the Cichlidae). Eggs are also borne externally in a number of exotically different ways in various families (reviewed by Breder & Rosen 1966, Balon 1975a). Most of the species are relatively small though *Osteoglossum* reaches a length of two meters.

Table 1. Families of fishes in which the species regularly exhibit parental care, or in which only some do, or none. Data in parentheses are the numbers of families.

Environment	Parental	Some parental	None parental	Sum
Freshwater[1]	41% (25)	13% (8)	46% (28)	(61)
Caribbean[2]	17% (10)	2% (1)	81% (49)	(60)
Marshall & Gilbert Is.[3]	12% (6)	6% (3)	82% (41)	(50)

[1]Loiselle (unpublished)
[2]Randall (1968)
[3]Schultz et al. (1961)

Internal fertilization with development in the reproductive tract has evolved in a number of small species in a variety of families, including the freshwater representatives, only, of the halfbeaks (Hemiramphidae) and within the Cyprinodontiformes, the Goodeidae, Poeciliidae, Tomeuridae, Jenynsiidae and Anablepidae (in part, from Breder & Rosen 1966, Lowe-McConnell 1975).

The foregoing supports the contention that postzygotic investment is more widespread, diverse and prolonged among tropical freshwater fishes than coral-reef fishes. It is more frequent among small than large species in freshwater, and that might be merely a consequence of there being more small species there; but the existence of a number of large species of freshwater fishes that guard young stands in contrast to their near absence among coral-reef fishes.

Why don't all coral-reef fishes broadcast pelagic eggs?

Small species are faced with a dilemma, if there is a need to disperse between reef systems. Most are so closely tied to the reef as to be compared to sessile invertebrates (Tyler 1971, Lassig 1977). And individuals are too small to produce enough propagules to have a reasonable chance of some successfully dispersing, since there is a limited range of optimal egg sizes.

How small species cope

Investment per offspring in caretaking species

Small species minimize the early mortality their offspring would experience before reaching a distant reef system in two ways: (1) they keep the eggs in a safe haven (see general discussion by Shine 1978), (2) they produce large young that are better able to avoid predators and to resist being dispersed into the open ocean (see below). This greater investment per egg further reduces fecundity in small species, with a trade-off between producing larger young and fewer eggs.

The evidence to test the hypothesis that small caretaking species in the tropics have relatively large eggs is nowhere conveniently summarized. Among marine coastal fishes of Great Britain, however, the data (Russell 1976) indicate that demersal (usually cared for) eggs tend to be larger than planktonic ones (Fig. 2). Only 16% (9 of 55) of the species with planktonic eggs have eggs as large or larger than 1.50 mm (17% in Fig. 1 of Ware 1975). In contrast, 48% (10 of 21) of the species that tend their demersal eggs have eggs larger than 1.50 mm. Demersal eggs are usually, but by no means always, cared for, and regularly have higher concentrations of respiratory pigments (Balon 1975b, 1978), presumably for anaerobic metabolism, and more yolk with higher concentrations of lipids (Balon 1977). The structure of adhesive eggs is also more complex than that of planktonic ones. Special vascular surfaces and associated structures are often elaborated to enhance gaseous exchange (Balon 1975b). This greater investment per offspring should further reduce fecundity in species that care for eggs deposited on the substrate.

Bearing young is but another form of postzygotic investment. Internally borne young are costly, whether carried in the reproductive tract, the mouth or in a pouch; they are often relatively large (Shaw 1971, Russell 1976).

The ophidioid fishes are of special interest here

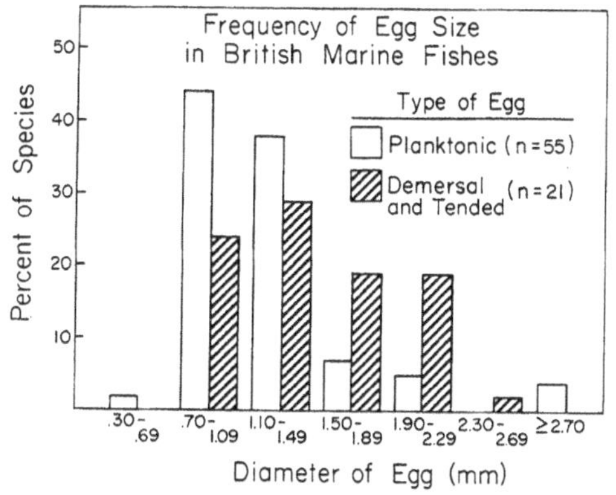

Fig. 2. Frequency of occurrence, expressed as percent of all species, of different sizes of eggs that are planktonic (open columns) and that are demersal (columns with diagonal lines) (data from Russell 1976).

because they are the most abundant and widely distributed ovoviviparous coral-reef fishes. Not much is known about them, but the pattern of investment per young, so far as known, is consistent with the theme of this essay (Table 2). Smaller species invest more per offspring. This is seen not only in the size of the young but also in the progression from ovoviviparity to viviparity. The trend should become clearer with data from more species, separated according to environment. The set of data shown in Table 2 are confounded by the fact that *Ogilbia cayorum* occurs in open-bottom habitats; its young are born highly developed (Saurez 1975); their behavior and morphology suggest a mixture of responses, perhaps alternating between pelagic and benthic habits, since they are essentially miniature adults and take cover, yet they are transparent (Longley & Hildebrand 1941).

These data (see also Table 3) indicate that small size of female is usually associated with relatively greater investment per egg, or per embryo in viviparous or ovoviviparous fishes. As a consequence,

Table 2. Size and number of young in relation to length of female in some shallow-water ophidioid fishes. All measurements in millimeters.

Species	♀ Length	No. eggs/larvae	Larvae length
Oligopus longhursti[1]	99	6,500	3.25
Dinematichthys iluceoteoides[2]	40–90	280–450	5.6–6.0
Ogilbia cayorum[3]	29–60	14 (max.)	12.0 +

[1]Wourms & Cohen (1975)
[2]Wourms & Bayne (1973)
[3]Suarez (1975)

Table 3. The relationship between mean body size, mean egg diameter, both in mm, and mean number of eggs in six species of cardinalfishes (Apogonidae) in Hawaii. Data from Chave (1971).

Species	SL	Egg ∅	No. of eggs
Apogonichthys waikiki			
sample a.	35	0.59	1,500
sample b.	43	0.61	5,100
Apogon erythrinus	43	0.49	2,600
Foa brachygramma	52	0.44	4,800
Apogon snyderi	88	0.41	20,000
Apogon maculiferus	97	0.43	17,000
Apogon menesemus	129	0.47	19,500

such females are even less fecund than their small size would predict. The investment in demersal eggs is commonly extended through the period of egg guarding, though now the cost is borne by the male. That is, in small species with external fertilization a female can compensate for her low fecundity by avoiding caretaking, leaving that to the male while she seeks food to produce more eggs.

From the foregoing one would expect larvae of small species with demersal eggs to occur close to the reef; those of large species with planktonic eggs should be found further offshore. Leis & Miller (1976), reviewing their plankton studies and those of others in Hawaii, found this to be true and cited a recent Russian study done off Cuba that produced a similar set of findings. I have replotted some of their data, eliminating the bay sample and adding larval eels they did not include (Fig. 3).

Prolonged larval existence

Fish larvae in general seem to be able to prolong their pelagic life, within limits, until they reach a

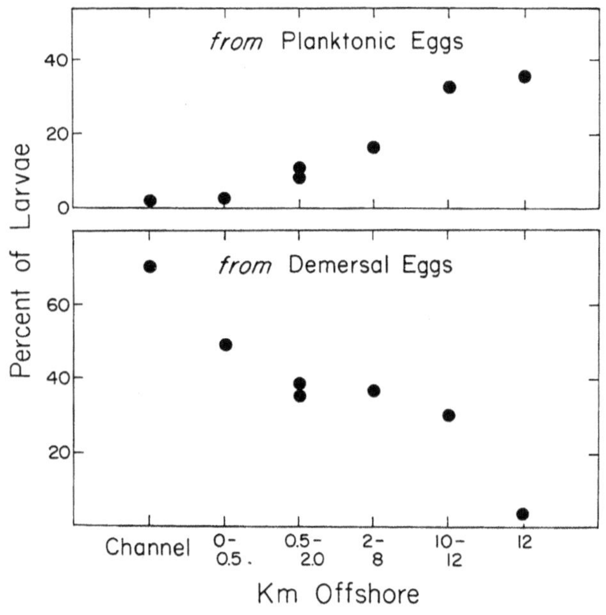

Fig. 3. The upper panel shows the percent of larvae of inshore coral-reef fishes that started out as planktonic eggs, the lower panel as demersal eggs. The abcissa gives the distance offshore at which the plankton were captured; 'Channel' refers to a pass through a reef into a bay. The data from 12-km collections were made only at the surface (after Leis & Miller 1976, who presented a similar figure.)

suitable substrate for settling out. Thorson (1950) first established this as a wide-spread trait among invertebrate larvae. Breder (1949) has documented the same phenomenon in some fish species with specialized larval stages (see also Randall 1961). The larvae of the apogonid *Foa brachygrammus* can settle at a length as small as 2 mm (1-10 days old) or as large as 22 mm (34 days) (Chave 1971). Gobies of the genus *Gillichthys* (not a coral-reef species) likewise show a wide range of sizes when they descend to the bottom (Barlow 1963).

As Rosenblatt (1963) pointed out, the tendency for small reef fishes to produce few larvae with a short pelagic phase restricts gene flow and leads to greater endemism (see also Rosenblatt & Walker 1963), and hence more speciation among the small fishes that invest postzygotically than among the large broadcasters. An extreme case is *Acanthochromis polyacanthus*, the damselfish that lacks a pelagic phase because of prolonged biparental care. Allen (1975) commented on the great variation in coloration between populations of *Acanthochromis:* 'It was almost as though a different color variety existed for every location we visited, even if separated by only 20 - 30 miles.' Genetic analysis indicates local endemism in *Acanthochromis* in contrast to relative uniformity between populations of *Dascyllus* (Ehrlich 1975), a damselfish whose young are initially planktonic.

The Starksiini, a New World tribe of predominantly tropical clinids with several species, mostly in the genus *Starksia*, is another example. 'Whereas other reef species are being shown to have widespread distributions throughout the tropical Western Atlantic, the species of *Starksia* exhibit distinct, restricted distributions' (Greenfield 1979). The distribution is not paradoxical in light of their biology. They are very small (20 – 60 mm SL), ovoviviparous species (Böhlke & Springer 1961, Rosenblatt & Taylor 1971). They produce few, relatively large young, e.g. a 28-mm *Xenomeda rhodopyga* contained about twenty-four 5 mm embryos; a 35 mm *S. spinipenis* had 121 embryos (Rosenblatt & Taylor 1971). Dispersal is most likely confined to a limited area (but see Jackson 1974).

Gobies of the genus *Gobiosoma* in the Caribbean Sea offer a clear parallel (Colin 1975). They are very small species that lay and take care of few large eggs. An analysis of the subgenus *Elacatinus* revealed a complex biogeographical pattern consisting of several species and commonly with indications of subdivisions within those species.

Seen in this light, there appears to be a positive correlation between size of a species and the size of its breeding stock. Therefore, genetic heterozygosity should be proportionate to size, i.e. small species (and large ones with limited dispersal) should be the least heterozygous. When the data on genetic variation within several species of fishes, presented by Somero & Soule (1974), are viewed in this way, they suggest that the relationship is worth exploring. A striking convergence exists in the similar situation of two sea stars (Echinodermata) from the North Pacific Coast of North America (Menge 1975).

Leptasterias hexactis is a small sea star, has a short life, broods a few large eggs, has low egg mortality, and the large young are not dispersed. *Pisaster ochraceus*, in contrast, is a large species with a long life that broadcasts vast numbers of planktonic eggs and disperses its small young widely. Menge suggested that because of high mortality to planktonic propagules and low lifetime fecundity, *Leptasterias* engages in high postzygotic investment at a cost to dispersal. The parallels with reef fishes are obvious.

Discussion

I have argued that the reproductive tactics of coral-reef fishes have been driven largely by selection for effective dispersal in a patchy environment. In doing so I have presented a simplified scenario, glossing over some thorny issues and minimizing exceptional cases. In the following I treat some of the difficulties and possible exceptions, though I do not claim to have resolved them all.

Why disperse?

Gadgil (1971) and Lidicker (1975) have pointed out that the significance of dispersal has been largely ignored in life history models. Waddington (1975) noted that the variant offspring of sexually reproducing animals should disperse to encounter new environments. Dispersal also reduces the probability of inbreeding (Bengtsson 1978), and minimizes the chances of density-dependent mortality and competition among one's offspring. Dispersal also spreads out the risk of mortality to offspring, if it is variable in space and time. On the other hand, dispersal inevitably increases the chance of mortality and carries the propagule away from the environment to which the parents are adapted (Thorson 1946).

There is general agreement that a fluctuating environment should favor dispersal (Wilson 1975). While opinions vary concerning the stability of coral reefs (Johannes 1978, Sale 1978b), I favor the proposition that reef instability has been an important factor selecting for dispersal, for reasons I have already outlined.

Apparently exceptional species

Differential survivorship of larvae. Within any broadly embracing scheme such as this there are apparent exceptions. I have considered only post-zygotic investment as compensation for high mortality during early ontogeny, and its costs to effective dispersal and colonization. Secure havens could be provided, however, by hiding eggs and/or producing toxic eggs.

While few eggs are known to be toxic (Orians & Janzen 1944), they have not been sought, especially in the tropics. Some tropical plectognaths lay toxic eggs (Halstead 1967), and other tropical groups, such as scorpaenids, grammistines, acanthurids, siganids and even gobiids produce toxins (Randall 1959, Halstead 1967, Randall et al. 1971, Lam 1974, Colin 1975).

Arguments about safe havens assume that early mortality for all species is basically the same, with survivorship improving with age (see examples for temperate-zone fishes in Marr 1956, Cushing 1974, May 1974, Lasker & Smigh 1977). But we have no knowledge of the survivorship curves for coral-reef species. The eggs of pteroids (Fishelson 1975) and antennariids (Breder & Rosen 1966) are released in gelatinous balls or ribbons that probably deter small predators. Some species have evolved highly effective planktonic forms that can persist in open water until remarkably large, as in the filefish *Pervagor melanocephalus* (Schultz et al. 1966) or the specialized larval damselfishes of the genus

Chromis, and salariine blennies (Leis & Miller 1976). Some forms (e.g. *Schindleria*) have evolved into pelagic paedomorphic species which reproduce in the open sea. Another is an undescribed paedomorphic goby apparently close to *Lythrypnus*; a female (14 mm SL) was taken full of eggs far offshore in the vicinity of the Galápagos Islands (Barlow unpublished data).

Rafting. Another possibility for dispersal, but with lower mortality, is rafting in flotsam. Flotsam forms foci for settling larvae, for mobile planktivores (notably several species of pomacentrids) and for a few benthic species that live on the irregular surface on the raft (e.g. *Blenniolus*) as though it were a mobile piece of reef (Hunter & Mitchell 1966). Occasional larvae might prolong their dispersal by settling out on a raft. Rafting is almost precluded as a means of dispersal for most adult coral-reef fishes because so little raft material emanates form the reef, other than for species that might associate intimately with drifting algal patches torn loose from the reef, such as antennariids, filefishes, pipefishes or seahorses.

Protogynous hermaphrodites. Many of the small species that broadcast eggs are protogynous hermaphrodites. An individual hermaphrodite can produce no more eggs than a female gonochorist, so dispersal per se is not improved. However, if two young protogynous hermaphrodites reach an unoccupied patch, one will likely become a male and they can reproduce. If they are gonochorists, there is only a 50% chance they would be of the opposite sex and so be able to reproduce (see also Tomlinson 1966, Charnov & Bull 1977).

Triggerfishes (Balistidae). Other fishes that might be considered exceptional are large species that invest postzygotically, except that none is known that lives in close association with the coral reef. The obvious group, based on information from studies not yet published, is the Balistidae (sensu stricto). Lythgoe & Lythgoe (1971) noted with little amplification that grey triggerfish *Balistes capriscus*, a subtropical species (about 50 cm SL), guards eggs for two days in a depression in the sand. Lobel (1978) mentioned that *Pseudobalistes flavimarginatus* builds nests. I observed a large triggerfish guarding a nest at Isla Cocos off Costa Rica. Recently, Hans Fricke (in litt.) has documented brief care of eggs by haremic females of two Red Sea species; in each case nests were in sand near a reef. The filefishes (Monacanthidae) are sometimes included as balistids; two species, *Monacanthus cirrhifer* and *Aleutera schoepfi*, have adhesive demersal eggs (Breder & Rosen 1966).

Two species of triggerfishes in the Gulf of California, *Balistes polylepis* and *Sufflamen verres* both nest in sand depressions at the reef's edge (Strand 1978). The male and female guard the eggs in *S. verres*, whereas the female is the sole caretaker in *B. polylepis* (which may be haremic). The eggs form a cohesive gelatinous mass in both species. The eggs are small, estimated at less than 1 mm diameter, and, most remarkably, female *B. polylepis* guard swimming young (Strand, personal communication). That large triggerfish invest postzygotically cannot yet be explained by the thesis of this aricle, but we need to know more about these fishes, relating the presence, or degree, of postzygotic investment to degree of association with the reef and the ability of adults to disperse.

Rabbitfishes (Siganidae). The siganids are large, mobile, gregarious fishes, living in groups from sizable schools down to a few individuals. These herbivores are widely distributed over both the reef and sandy areas. They apparently move in vast schools onto the reef flat toward evening with the rising tide, and spawn in the middle of the night at high tide. Their spherical (0.6 mm diameter) eggs are demersal and adhesive; apparently they lie and adhere where they fall (Lam 1974). By spawning enormous numbers of eggs (300,000 to 500,000 per female) synchronously in a given area, on a relatively seasonal basis, they probably swamp local predators on both eggs and young.

Sharks. While this essay has been limited to teleosts, I can briefly consider some other groups of fishes. Sharks are large species, most of which are ovoviviparous or viviparous, and strong swimmers that should be able to disperse across open water. Even the sluggish *Triaenodon obesus* has crossed the East Pacific Barrier (Rosenblatt et al. 1972). One pregnant female suffices for colonization. This ability to disperse as adults and colonize new reefs probably accounts for why there are so few species of sharks, and why they are so widely distributed. Small bottom dwelling sharks (often oviparous) are conspicuously absent from coral atolls.

Coelacanths. The coelacanth *Latimeria chalumnae* deserves mentioning as a large species, closely tied to the bottom and thus poorly adapted to life in open water. Reproduction (Smith et al. 1975) is through internal fertilization of huge eggs (85 – 90 mm diameter) which develop ovoviviparously into juveniles (about 30% adult size) before birth (Balon 1977). Since *Latimeria* is confined to depths between 70 and 600 m (McCosker 1979), dispersal between islands other than those connected by shallow submarine ridges is ruled out. It is no surprise that this species is confined to a few islands of the oceanic Comoran Archipelago.

Why don't large coral-reef fishes invest postzygotically?

All else being equal, parental care of eggs or viviparity would increase individual fitness through the higher probability of survival of more offspring. A large female should produce as many eggs over her lifetime, whether spawned frequently in small numbers as now happens, or in larger numbers separated by intervals of male parental care, as in smaller species. All else must not be equal. Postzygotic investment must reduce lifetime fecundity (Trivers 1972, Maynard Smith 1977) or constrain dispersal. Some of the costs are obvious, other are not.

Female postzygotic investment. If the female retains the eggs, even if she no longer contributes to their nutrition (ovoviviparity), she cannot develop another clutch until the present one is released. Likewise, a female actively caring for demersal eggs will be unable to feed as well and therefore cannot develop another clutch as quickly as a non-caretaking female. In either case, viviparous or caretaking females would produce fewer eggs over their lifetime than ones that feed continuously and broadcast a steady production of eggs. Viviparity and female caretaking should be rare at any size, and they are.

Males as a limiting resource. If male caretakers are not always available to females ready to spawn, those females, and those males' daughters, will have lowered life-time fecundities. Males will compete for the (usually) limited nest sites suitable for caring for eggs (Loiselle 1978, Baylis 1981). Only a few males will be successful, and although most may be polygynous the number of clutches each can accommodate may be constrained by the nature of the site, phasing of caretaking to development of the embryos, or his ability to protect a growing mass of eggs.

For these reasons, males might become a limiting resource to females (see Mashiko 1976 for a possible example in the Eleotridae). Males might be limiting in terms of frequency of breeding, compounding the limitations of suitable nesting sites. If so, there should be competition among females for males (Trivers 1972). That males are usually in limited supply has not been reported and may be uncommon in substrate brooders, but it might be common in male bearers.

Risk of predation on caretaker males. Caretaking probably increases the risk of predation on the male himself. However, his mortality should not limit the females' opportunities to breed if males are polygynous and if an excluded male takes over the breeding territory. As yet there is no estimate of the relative risks of caretaking versus migrating to the edge of the reef to spawn in an upward rush.

Cost of cared-for eggs. Demersal eggs are generally larger, structurally more complex and richer biochemically, with a larger proportion of ancillary structures and take a longer time until the young begin exogenous feeding. Even leaving aside the costs of caretaking, having demersal eggs is usually associated with some reduction in fecundity.

Problems facing larvae. To move from the nest to the surface the young must be strong swimmers. As such they require a longer period of development, hence greater investment per young and so reduced fecundity. The young may risk encounters with sessile filter feeders (Fricke 1976), which are seldom a problem for pelagic eggs and young. Finally, the young start their voyage from the nest, not the edge of the reef. The sum of these risks is real but difficult to compare to the risks in broadcasting eggs from the edge of the reef.

Ability to defend small young. If eggs and young are to be numerous enough for a successful dispersal strategy, they must be small. Therefore predators on them will be much smaller than parents. Very large fish may not be sufficiently manoeuvrable to defend mobile young from predators (a stationary egg is much easier to defend). Perhaps species such as the triggerfish care for eggs and young in the few places near the edge of the reef where small predators are rare. If a large fish produced young sufficently large to defend easily, it would have to do so through huge eggs or viviparity. In either case fecundity would be reduced, and young would likely be nonpelagic, with a loss in ability to disperse, as in *Latimeria*.

Advantages in dispersing as an egg. Starting dispersal as an egg (or embryo) instead of as a larva (or juvenile) could be advantageous in that most of the journey would be completed with endogenous nutrition. A representative time to initiation of exogenous feeding for a coral-reef fish is about 7 days. In that time, with a typical current of about 5 km h^{-1}, a nonfeeding propagule could drift about 800 km, which is often enough to reach another reef system.

Closing comments

Parental care among cora-reef fishes cannot be discussed without regard to the size of fish. Caretaking predominates among small species and is virtually nonexistent among large ones. But in all of them, postzygotic investment is minimal compared to freshwater species because of the pervasive need to disperse as eggs or young. The situation on coral reefs therefore presents a specific problem whose solution offers some general principles (Menge 1975).

Postzygotic investment may be expected in general when (1) high mortaility of the young stages occurs in the absence of some form of parental care (the hostile environment of Thorson 1950), and (2) the fecundity of the parent is limited. A third condition (3) the possibility of adult dispersal, facilitates postzygotic investment.

Fecundity might be constrained by other than small size, for instance a short breeding season. That could explain the higher frequency of parental care in marine invertebrates (Thorson 1950) and fishes at high latitudes. That could also be involved in the higher frequency of caretaking in freshwater fishes. They often experience seasonality, even in the tropics, restricting reproduction to a limited period. Interestingly, both in cold marine and in freshwater environments, caretaking is also found in some large species of fishes.

There should always be a trade-off between dispersal as an adult and investment per offspring. In coastal fishes, and especially in freshwater fishes, adults have more opportunity to disperse. When that is the case, larger investment per offspring, including caretaking, is possible.

This essay is also relevant to such an apparently unrelated situation as the effects of nuclear explosions on coral-reef communities, such as at Enewetak Atoll, where I have made many observations. There is little if any perceptible damage to the community away from the blast site. Mutations should be most evident in the pelagic propagules. But these are borne away from the home reef, especially those of the larger species. Since the young of small species are more likely to settle within the home reef system, mutants that survive to adulthood are more likely to be found among small species. Yet, because they are inconspicuous and commercially unimportant, the small species tend to be ignored.

Acknowledgements

A nascent and more broadly ranging version of this paper was presented as an invited paper in a conference on fish sociobiology, Seewiesen, Germany in September, 1977, and a condensed version was given in the Symposium on Natural Selection and Social Behavior, Ann Arbor, Michigan, in October, 1978. After that, I narrowed the scope to coral-reef fishes and presented this essay to the

International Ethological Conference, Vancouver, B.C., in August, 1979, and to the Second Biennial Conference for the Ethology and Behavioral Ecology of Fishes, Normal, Illinois, in October, 1979. On each occasion I profited greatly from the discussions that followed; my thanks go to those discussers.

This article has also benefitted enormously from the candid criticisms of a number of readers, to whom I am truly indebted: C. G. Barnett, A. W. Ebeling, E. A. Fischer, S. G. Hoffman, T. M. Lim, P. V. Loiselle, D. L. G. Noakes, W. Rogers, R. H. Rosenblatt, R. R. Warner and G. C. Williams. They made numerous suggestions for the improvement of this essay, ranging from syntax through organization, logic and documentation. As only good friends may do, they took delight in misunderstanding each and every ambiguity, no matter how slight; doubtless, new ones have crept into the revision. There remain areas of disagreement, so the final responsibility for the content of the article rests with me.

I am also grateful to the Department of Energy for supporting the field work at Enewetak Atoll, through the good offices of the Mid-Pacific Marine Laboratory of the University of Hawaii and its capable director Ernst S. Reese and his colleague Victor Johnson. It gives me pleasure to acknowledge the long-standing support of the National Science Foundation, including the writing of this essay (NSF Grant BNS 7680115) and the funding in the past of field work in Nicaragua, Panama and British Honduras. While I have dived in many other waters, marine and fresh, the opportunity for intensive observation provided by that support sowed the seeds for this article.

References cited

Abbot, D. P. 1966. Factors influencing the zoogeographic affinities of the Galápagos inshore marine fauna. pp. 108–122. *In*: R. I. Bowman (ed.) The Galápagos, University of California Press, Berkeley.

Adey, W. H. 1978. Coral reef morphogenesis: a multidimensional model. Science 202: 831–837.

Allen, G. R. 1972. The anemonefishes: their classification and biology. T. F. H. Publications, Neptune City. 272 pp.

Allen, G. R. 1975. Damselfishes of the South Seas. T. F. H. Publications, Neptune City. 240 pp.

Bagenal, T. B. 1967. A short review of fish fecundity. pp. 89–111. *In*: S. D. Gerking (ed.) The Biological Basis of Freshwater Fish Production, I. B. P. Sectional Committee on Productivity of Freshwater Communities. Blackwell, Oxford.

Bagenal, T. B. 1971. The inter-relations of the size of fish eggs, the date of spawning and the production cycle. J. Fish Biol. 3: 207–219.

Balon, E. K. 1975a. Reproductive guilds of fishes: a proposal and definition. J. Fish. Res. Board Can. 32: 821–864.

Balon, E.K. 1975b. Terminology of intervals in fish development. J. Fish. Res. Board Can. 32: 1663–1670.

Balon, E. K. 1977. Early ontogeny of *Labeotropheus* Ahl, 1927 (mbuna, Cichlidae, Lake Malawi), with a discussion on advanced protective styles in fish reproduction and development. Env. Biol. Fish. 2: 147–176.

Balon, E. K. 1978. Reproductive guilds and the ultimate structure of fish taxocenes: amended contribution to the discussion presented at the mini-symposium. Env. Biol. Fish. 3: 149–152.

Barlow, G. W. 1963. Species structure of the gobiid fish *Gillichthys mirabilis* from coastal sloughs of the Eastern Pacific. Pacif. Sci. 17: 47–72.

Barlow, G. W. 1964. Ethology of the Asian teleost *Badis badis*. V. Dynamics of fanning and other parental activities, with comments on the behavior of the larvae and postlarvae. Z. Tierpsychol. 21: 99–123.

Barlow, G. W. 1974. Contrasts in social behavior between Central American cichlid fishes and coral-reef surgeon fishes. Amer. Zool 14: 7–32.

Barlow, G. W. 1975a. On the sociobiology of four Puerto Rican parrotfishes (Scaridae). Mar. Biol. 33: 281–293.

Barlow, G. W. 1975b. On the sociobiology of some hermaphroditic serranid fishes, the hamlets, in Puerto Rico. Mar. Biol. 33: 295–300.

Baylis, J. R. 1981. The evolution of parental care in fishes, with reference to Darwin's rule of male sexual selection. Env. Biol. Fish. 6 (in print).

Beamish, F. W. H. 1978. Swimming capacity. pp. 101–187. *In*: W. S. Hoar & D. J. Randall (ed.) Physiology of Fishes, Vol. 7, Academic Press, New York.

Bengtsson, B. O. 1978. Avoiding inbreeding: at what cost? J. Theoret. Biol. 73: 439–444.

Blache, J. 1964. Les poissons du bassin du Tchad et du bassin adjacent de Mayo Kebbi. Office de la Recherche Scientifique et Technique Outre-Mer, Paris, 483 pp.

Blaxter, J. H. S. 1969. Development: eggs and larvae. pp. 177–252. *In*: W. S. Hoar & D. J. Randall (ed.) Fish Physiology, Vol. 3, Academic Press, New York.

Böhlke, J. E. & C. C. G. Chaplin. 1957. Oral incubation in Bahaman jawfishes *Opisthognathus whitehursti* and *O. maxillosus*. Science 125: 353.

Böhlke, J. E. & C. C. G. Chaplin. 1968. Fishes of the Bahamas and adjacent tropical waters. Livingston, Wynnewood. 771 pp.

Böhlke, J. E. & V. G. Springer. 1961. A review of the Atlantic species of the clinid fish genus *Starksia*. Proc. Acad. Nat. Sci. Philad. 113: 29–60.

Böhlke, J. E. & L. P. Thomas. 1961. Notes on the west Atlantic jawfishes, *Opisthognathus aurifrons*, *O. lonchurus*, and *Gnathypops bermudezi*. Bull. Mar. Sci. Gulf Carib. 11: 503–516.

Bortone, S. A. 1976. Effects of a hurricane on the fish fauna of Destin, Florida. Florida Sci. 39: 245–248.

Breder, C. M. 1941. On the reproductive behavior of the sponge

blenny, *Paraclinus marmoratus* (Steindachner).Zoologica 26: 233–235.
Breder, C. M. 1949. On the taxonomy and the postlarval stages of the surgeonfish, *Acanthurus hepatus*. Copeia 1949: 296.
Breder, C. M. 1962. On the significance of transparency in osteichthid fish eggs and larvae. Copeia 1962: 561–567.
Breder, C. M. & D. E. Rosen. 1966. Modes of reproduction in fishes. Natural History Press, Garden City, New York. 941 pp.
Bryant, G. L. 1977. Fecundity and growth of the tule perch, *Hysterocarpus traski*, in the lower Sacramento-San Joaquin delta. Calif. Fish Game 63: 140–156.
Burgess, G. H. & C. R. Gilbert. 1978. Fishes of Isla de Providencia, Columbia. A. S. B. Bull. 25: 1–58.
Casteel, R. W. 1978. *Damalichthys saratogensis*, a new freshwater fish (Embiotocidae) from the Plio-Pleistocene (Blancan) of California. Copeia 1978: 293–298.
Charney, P. 1976. Oral brooding in the cardinalfishes *Phaeoptyx conklini* and *Apogon maculatus* from the Bahamas. Copeia 1971: 198–200.
Charnov, E. L. & J. Bull. 1977. When is sex environmentally determined? Nature 266: 828–830.
Chave, E. H. 1971. Ecological requirements of six species of cardinalfishes (Genus *Apogon*) in small geographical areas of Hawaii. Ph. D. Thesis, University of Hawaii, Honolulu. 279 pp.
Choat, J. H. & D. R. Robertson. 1975. Protogynous hermaphroditism in fishes of the family Serranidae. pp. 263–283. *In*: R. Reinboth (ed.) Intersexuality in the Animal Kingdom, Springer Verlag, New York.
Clark, E., A. Ben-Tuvia & H. Steinitz. 1962. Observations on a coastal community, Dahlak Archipelago, Red Sea. Israel South Red Sea Exped., 1962, Report 30: 15–31.
Clark, E. & H. A. F. Gohar. 1953. The fishes of the Red Sea: Order Plectognathi. Publ. Mar. Biol. Stat. Al Ghardaqa (Red Sea) 8: 1–80.
Colin, P. L. 1972. Daily activity patterns and effects of environmental conditions on the behavior of the yellowhead jawfish, *Opistognathus aurifrons*, with notes on its ecology. Zoologica 57: 137–169.
Colin, P. 1975. The neon gobies: The comparative biology of the gobies of the genus *Gobiosoma*, subgenus *Elacatinus* (Pisces: Gobiidea), in the tropical Western North Atlantic Ocean. T. F. H. Publications, Hong Kong. 304 pp.
Colin, P. L. 1978. Daily and summer-winter variation in mass spawning of the striped parrotfish, *Scarus croicensis*. U. S. Fish. Bull. 76: 117–124.
Colin, P. L. & D. W. Arneson. 1978. Aspects of the natural history of swordtail jawfish, *Lonchopisthus micrognathus* (Poey) (Pisces: Opistognathidae), in south-western Puerto Rico. J. Nat. Hist. 12: 689–697.
Colin, P. L. & H. E. Clavijo. 1978. Mass spawning by the spotted goatfish, *Pseudupeneus maculatus* (Bloch) (Pisces: Mullidae). Bull. Mar. Sci. 28: 780–782.
Cushing, D. H. 1974. The possible density-dependence of larval mortality and adult mortality in fishes. pp. 103–111. *In*: J. H. S. Blaxter (ed.) The Early Life History of Fish, Springer Verlag, New York.
Dawson, C. E. 1971. Supplemental observations on *Gunterichthys longipenis*, a northern Gulf of Mexico brotulid fish. Copeia 1971: 164–167.
Ehrlich, P. R. 1975. The population biology of coral reef fishes. Ann. Rev. Ecol. System. 6: 211–247.
Ehrlich, P. R., F. H. Talbot, B. C. Russel & G. R. V. Anderson. 1977. The behaviour of chaetodontid fishes with special reference to Lorenz's 'poster colouration' hypothesis. J. Zool. (London) 183: 213–228.
Fischer, E. A. 1979. Mating system and simultaneous hermaphroditism in the coral reef fish, *Hypoplectrus nigricans* (Serranidae), with a discussion of the systematic status of the species in the genus *Hypoplectrus*. Ph.D. Thesis, University of California, Berkeley. 156 pp.
Fishelson, L. 1966. *Solenostomus cyanopterus* Bleker [sic] (Teleostei, Solenostomidae) in Eilat (Gulf of Akaba). Israel J. Zool. 15: 95–105.
Fishelson, L. 1975. Ethology and reproduction of pteroid fishes found in the Gulf of Aqaba (Red Sea), especially *Dendrochirus brachypterus* (Cuvier), (Pteroidae, Teleostei). Pubbl. Staz. Zool. Napoli Suppl. 39: 635–656.
Fishelson, L., D. Popper & A. Avidor. 1974. Biosociology and ecology of pomacentrid fishes around the Sinai Peninsula (northen Red Sea). J. Fish Biol. 6: 119–133.
Fricke, H. W. 1976. Bericht aus dem Riff. Piper, Munich. 254 pp.
Gadgil, M. 1971. Dispersal: population consequences and evolution. Ecology 52: 253–261.
Gill, T. 1907. Life histories of toadfishes (batrachoids), compared with those of weevers (trachinids) and stargazers (uranoscopids). Smithsonian Misc. Coll. 48: 388–427.
Gooding, R. M. & J..J. Magnuson. 1967. Ecological significance of a drifting object to pelagic fishes. Pacif. Sci. 21: 486–497.
Goreau, T. F. & W. D. Hartman. 1963. Boring sponges as controlling factors in the formation and maintenance of coral reefs. pp. 25–54. *In*: R. F. Sognnaes (ed.) Mechanisms of Hard Tissue Destruction, Amer. Assoc. Advan. Sci., Washington, D. C.
Greenfield, D. W. 1979. A review of the Western Atlantic *Starksia ocellata*-complex (Pisces: Clinidae) with the description of two new species and proposal of superspecies status. Fieldiana Zool. 73: 9–48.
Gudger, E. W. 1910. Habits and life history of the toadfish *Opsanus tau*. Bull. U. S. Bur. Fish. 28: 1071–1109.
Gunter, G. 1947. Observations on breeding of the marine catfish, *Galeichthys felis* (Linnaeus). Copeia 1947: 217–223.
Haldane, J. B. S. 1928. Possible worlds. Chatto & Windus, London, 225 pp.
Halstead, B. W. 1967. Poisonous and venomous marine animals of the world. Vol. 2. Vertebrates. U. S. Government Printing Office, Washingon, D. C. 1070 pp.
Helfrich, P. & P. M. Allen. 1975. Observations on the spawning of mullet *Crenimugil crenilabis* (Forskål) at Enewetak, Marshall Islands. Micronesica 11: 219–225.
Hiemstra, W. H. 1962. A correlation table as an aid for identifying pelagic fish eggs in plankton samples. J. Cons. Perm. Int. Explor. Mer. 27: 100–108.
Hildebrand, S. F. & L. E. Cable. 1934. Reproduction and development of whitings, or kingfishes, drums, spot, croaker and weakfishes or seatrouts, family Sciaenidae, of the Atlantic coast of the United States. Bull. U. S. Bur. Fish. 48: 41–117.
Hobson, E. S. 1965. Diurnal-nocturnal activity of some inshore fishes in the Gulf of California. Copeia 1965: 291–302.
Hobson, E. S. & J. R. Chess. 1978. Trophic relationships among fishes and plankton in the lagoon at Enewetak Atoll, Marshall Islands. U.S. Fish. Bull. 76: 133–153.
Holzberg, S. 1973. Beobachtungen zur Ökologie und zum Sozialverhalten des Korallenbarsches *Dascyllus marginatus* Rüppell (Pisces: Pomacentridae). Z. Tierpsychol. 33: 493–513.

Hubbs, C. L. 1948. Changes in the fish fauna of western North America correlated with changes in ocean temperature. Sears Found. J. Mar. Res. 7: 459–482.

Hunter, J. R. 1976. Report of a colloquium on larval fish mortality studies and their relation to fishery research, January 1975. NOAA Tech. Rept. NMFS CIRC-3: 1–5.

Hunter, J. R. & C. T. Mitchell. 1966. Association of fishes with flotsam in offshore waters of Central America. U. S. Fish. Bull. 66: 13–29.

Jackson, J. B. C. 1974. Biogeographic conseqences of eurytopy and stenotopy among marine bivalves and their evolutionary significance. Amer. Natur. 108: 541–650.

Johannes, R. E. 1978. Reproductive strategies of coastal marine fishes in the tropics. Env. Biol. Fish. 3: 65–84.

Jones, R. & W. B. Hall. 1974. Some observations on the population dynamics of the larval stage in the common gadoids. pp. 87–102. *In*: J. S. Blaxter (ed.) The Early Life History of Fish. Springer Verlag, New York.

Jones, R. S. 1968. Ecological relationships in Hawaiian and Johnston Island Acanthuridae (surgeonfishes). Micronesica 4: 309–361.

Keenleyside, M. H. A. 1972. The behaviour of *Abudefduf zonatus* (Pisces: Pomacentridae) at Heron Island, Great Barrier Reef. Anim. Behav. 20: 763–775.

Kerstitch, A. 1979. Blue-spotted jawfish. Freshw. Mar. Aquar. 9: 9–10.

Lam, T. J. 1974. Siganids: their biology and mariculture potential. Aquaculture 3: 325–354.

Lasker, R. & P. E. Smith. 1977. Estimation of the effects of environmental variation on the eggs and larvae of the northern anchovy. Calif. Cooper. Ocean. Fish. Invest. Rept. 19: 128–137.

Lassig, B. 1976. Field observations of the reproductive behaviour of *Paragobiodon* spp. (Osteichthyes: Gobiidae) at Heron Island, Great Barrier Reef. Mar. Behav. Physiol. 3: 283–293.

Lassig, B. R. 1977. Socioecological strategies adopted by obligate coral-dwelling fishes. Proc. Third Intern. Coral Reef Symp. 1: 565–570.

Leis, J. M. & J. M. Miller. 1976. Offshore distribution patterns of Hawaiian fish larvae. Mar. Biol. 36: 359–367.

Leong, D. 1967. Breeding and territorial behavior in *Opisthognathus aurifrons* (Opisthognathidae). Die Naturwissens. 54: 97.

Lidicker, W. Z., Jr. 1975. The role of dispersal in the demography of small mammals. pp. 103–134. *In*: International Biological Programme. Small mammals: Their productivity and population dynamics, Cambridge University Press, Cambridge.

Lindsey, C. C. 1966. Body sizes of poikilotherm vertebrates at different latitudes. Evolution 20: 456–465.

Lobel, P. S. 1978. Diel, lunar, and seasonal periodicity in the reproductive behavior of the pomacanthid fish, *Centropyge potteri*, and some other reef fishes in Hawaii. Pacif. Sci. 32: 193–207.

Loiselle, P. V. 1978. Prevalence of male brood care in teleosts. Nature 275: 98.

Loiselle, P. V. & G. W. Barlow. 1978. Do fishes lek like birds? pp. 31–75. *In*: E. S. Reese & F. J. Lighter (ed.) Contrasts in Behavior, Wiley, New York.

Longley, W. H. & S. F Hildebrand. 1941. Systematic catalogue of the fishes of Tortugas, Florida. With observations on color, habits, and local distribution. Carnegie Inst. Washington Publ. 535: 1–331.

Lowe-McConnell, R. H. 1975. Fish communities in tropical freshwaters. Longman, New York. 337 pp.

Luckhurst, B. E. & K. Luckhurst. 1977. Recruitment patterns of coral reef fishes on the fringing reef of Curacao, Netherland Antilles. Can. J. Zool. 55: 681–703.

Lüling, K. H. 1964. Zur Biologie und Ökologie von *Arapaima gigas* (Pisces: Osteoglossidae). Z. Morphol. Ökol. Tiere 54: 436–530.

Lythgoe, J. & G. Lythgoe. 1971. Fishes of the sea; the coastal waters of the British Isles, northern Europe, and the Mediterranean. Blackford, London. 320 pp.

Marr, J. C. 1956. The critical period in the early life history of marine fishes. J. Cons. Perm. Inter. Explor. Mer. 21: 160–170.

Marshall, N. B. 1953. Egg size in Arctic, Antarctic and deep-sea fishes. Evolution 7: 328–341.

Marshall, N. B. 1971. Explorations in the life of fishes. Harvard University Press, Cambridge. 204 pp.

Mashiko, K. 1976. Ecological study on breeding of an eleotrid goby, *Odontobutis obscurus* (Temminck et Schlegel), under rearing conditions. Jap. J. Ecol. 26: 91–100.

May, R. C. 1974. Larval mortality in marine fishes and the critical period concept. pp. 3–19. *In*: J. H. S. Blaxter (ed.) The Early Life History of Fish, Springer Verlag, New York.

Maynard Smith, J. 1977. Parental investment: a prospective analysis. Anim. Behav. 25: 1–9.

McCosker, J. E. 1979. Inferred natural history of the living coelacanth. pp. 17–24. *In*: J. E. McCosker & M. D. Lagios (ed.) The Biology and Physiology of the Living Coelacath. Occasional Papers of the California Academy of Sciences (134), San Francisco.

Menge, B. A. 1975. Brood or broadcast? The adaptive significance of different reproductive strategies in two intertidal sea stars *Leptasterias hexactis* and *Pisaster ochraceus*. Mar. Biol. 37: 87–100.

Meyer, K. A. 1977. Reproductive behavior and patterns of sexuality in the Japanese labrid fish *Thalassoma cupido*. Jap. J. Ichthyol. 24: 101–112.

Molles, M. C., Jr. 1978. Fish species diversity on model and natural reef patches: experimental insular biogeography. Ecol. Monogr. 48: 289–305.

Moyer, J. T. & L. J. Bell. 1976. Reproductive behavior of the anemonefish *Amphiprion clarkii* at Jiyake-jima, Japan. Jap. J. Ichthyol. 23: 23–32.

Moyer, J. T. & A. Nakazono. 1978a. Population structure, reproductive behavior and protogynous hermaphroditism in the angelfish *Centropyge interruptus* at Miyake-jima, Japan. Jap. J. Ichthyol. 25: 25–39.

Moyer, J. T. & A. Nakazono. 1978b. Protandrous hermaphroditism in six species of the anemonefish genus *Amphiprion* in Japan. Jap. J. Ichthyol. 25: 101–106.

Munroe, J. L., V. C. Gant, R. Thompson & R. H. Reeson. 1973. The spawning season of Caribbean reef fishes. J. Fish Biol. 5: 69–84.

Nikolsky, G. V. 1963. The ecology of fishes. Academic Press, New York. 352 pp.

Ogden, J. C. & P. R. Ehrlich. 1977. The behavior of heterotypic resting schools of juvenile grunts (Pomadasyidae). Mar. Biol. 33: 273–280.

Oppenheimer, J. R. 1970. Mouthbreeding in fishes. Anim. Behav. 18: 493–503.

Orians, G. H. & D. H. Janzen. 1974. Why are embryos so tasty? Amer. Natur. 108: 581–592.

Penrith, M.-L. 1969. The systematics of the fishes of the family

Clinidae in South Africa. Ann. So. Afr. Mus. 55: 1–121.
Pope, J. A., D. H. Mills & W. M. Shearer. 1961. The fecundity of the Atlantic salmon (*Salmo salar* Linn.). Freshw. Salmon Fish. Res. 26: 1–12.
Powles, H. 1978. Observations on benthic larvae of *Pareques* (Pisces: Sciaenidae) from Florida and Columbia. Copeia 1978: 169–172.
Qasim, S. Z. 1956. Time and duration of the spawning season in some marine teleosts in relation to their distribution. J. Cons. Perm. Intern. Explor. Mer. 21: 144–155.
Randall, J. E. 1959. Report of a caudal-spine wound from the surgeonfish *Acanthurus lineatus* in the Society Islands. Wassmann J. Biol. 17: 245–248.
Randall, J. E. 1961. A contribution to the biology of the convict surgeonfish of the Hawaiian Islands, *Acanthurus triostegus sandvicensis*. Pacif. Sci. 15: 215–272.
Randall, J. E. 1963. An analysis of the fish populations of artificial and natural reefs in the Virgin Islands. Carib. J. Sci. 3: 1–16.
Randall, J. E. 1968. Caribbean reef fishes. T. F. H. Publications, Neptune City. 318 pp.
Randall, J. E. 1974. The effects of fishes on coral reefs. Proc. 2nd Intern. Coral Reef Symp., Great Barrier Reef Comm., Brisbane 1: 159–166.
Randall, J. E., K. Aida, T. Hibiya, N. Mitsuura, H. Kamiya & Y. Hashimoto. 1971. Grammistin, the skin toxin of soapfishes, and its significance in the classification of the Grammistidae. Publ. Seto Mar. Biol. Lab. 19: 157–190.
Randall, J. E. & R. K. Kanayama. 1972. Hawaiian fish immigrants. Sea Frontiers 18: 144–153.
Randall, J. E. & H. A. Randall. 1963. The spawning and early development of the Atlantic parrotfish, *Sparisoma rubripinne*, with notes on other scarid and labrid fishes. Zoologica 48: 49–60.
Reese, E. S. 1964. Ethology and marine zoology. Oceanogr. Mar. Biol. Ann. Rev. 1964: 455–488.
Reese, E. S. 1975. A comparative field study of the social behavior and related ecology of reef fishes of the family Chaetodontidae. Z. Tierpsychol. 37: 37–61.
Robertson, D. R. 1973. Field observations on the reproductive behaviour of a pomacentrid fish, *Acanthochromis polyacanthus*. Z. Tierpsychol. 32: 319–324.
Robertson, D. R. & S. G. Hoffman. 1977. The roles of female mate choice and predation in the mating systems of some tropical labroid fishes. Z. Tierpsychol. 45: 298–320.
Robertson, D. R., N. V. C. Polunin & K. Leighton. 1979. The behavioral ecology of three Indian Ocean surgeonfishes (*Acanthurus lineatus*, *A. leucosternon* and *Zebrasoma scopas*): their feeding strategies, and social and mating systems. Env. Biol. Fish. 4: 125–170.
Robins, C. R. 1957. Effects of storms on the shallow water fish fauna of southern Florida with new records of fish from Florida. Bull. Mar. Sci. Gulf Carib. 7: 266–275.
Robins, C. R. 1971. Distributional patterns of fishes from coastal and shelf waters of the tropical Western Atlantic. pp. 249–255. Symp. Invest. Resour. Carib. Sea Adjacent Regions, Pap. Fish. Res. F. A. O., Rome
Rosenblatt, R. H. 1963. Some aspects of speciation in marine shore fishes. Syst. Assoc. Publ. 5: 171–180.
Rosenblatt, R. H., J. E. McCosker & I. Rubinoff. 1972. Indo-west Pacific fishes from the Gulf of Chiriqui, Panama. Nat. Hist. Mus. Los Angeles Contr. Sci. 234: 1–18.
Rosenblatt, R. H. & L. R. Taylor, Jr. 1971. The Pacific species of the clinid fish tribe Starksiini. Pacif. Sci. 25: 436–463.
Rosenblatt, R. H. & B. W. Walker. 1963. The marine shore-fishes of the Galápagos Islands. Calif. Acad. Sci. Occ. Pap. 44: 97–106.
Ross, R. M. 1978. Reproductive behavior of the anemonefish *Amphiprion melanopus* on Guam. Copeia 1978: 103–107.
Russell, F. S. 1976. The eggs and planktonic stages of British marine fishes. Academic Press, New York. 524 pp.
Sale, P. F. 1970. Distribution of larval Acanthuridae off Hawaii. Copeia 1970: 765–766.
Sale, P. F. 1977. Maintenance of high diversity in coral reef fish communities. Amer. Natur. 111: 337–359.
Sale, P. F. 1978a. Reef fishes and other vertebrates: a comparison of social structures. pp. 313–346. *In*: E. S. Reese & F. J. Lighter (ed.) Contrasts in Behavior, Wiley-Interscience, New York.
Sale, P. A. 1978b. Coexistence of coral reef fishes -- a lottery for living space. Env. Biol. Fish. 3: 85–102.
Schultz, L. P. 1948. The biology of Bikini Atoll with special reference to the fishes. Ann. Rept. Board Regents Smithson. Inst. 1947: 301–316.
Schultz, L. P., L. P. Woods & E. A. Lachner. 1966. Fishes of the Marshall and Marianas Islands. Smithsonian Inst. U. S. Nat. Mus. Bull. 202, 3 vol. 1299 pp.
Shaw, E. 1971. Natural precocial sexual maturity in the viviparous perch, *Cymatogaster aggregata* Gibbons (Embiotocidae, Perciformes). Amer. Mus. Novit. 2479: 1–10.
Shine, R. 1978. Propagule size and parental care: the 'safe harbor' hypothesis. J. Theoret. Biol. 75: 417–424.
Smith, A. C. 1969. Protein variation in the eye lens nucleus of the mackerel scad (*Decapterus bipinnulatus*). Comp. Biochem. Physiol. 28: 1161–1168.
Smith, C. L. 1975. The evolution of hermaphroditism in fishes. pp. 295–310. *In*: R. Reinboth (ed.) Intersexuality in the Animal Kingdom, Springer Verlag, New York.
Smith, C. L., C. S. Rand, B. Schaeffer & J. W. Atz. 1975. *Latimeria*, the living coelacanth, is ovoviviparous. Science 190: 1105–1106.
Somero, G. & M. Soulé. 1974. Genetic variation in marine fishes as a test of the niche-variation hypothesis. Nature 249: 670–672.
Springer, V. G. & A. J. McErlean. 1962. A study of the behavior of some tagged South Florida coral reef fishes. Amer. Midl. Natur. 67: 386–397.
Stearns, S. C. 1976. Life-history tactics: a review of the ideas. Quart. Rev. Biol. 51: 3–47.
Strand, S. W. 1978. Community structure among reef fish in the Gulf of California: the use of reef space and interspecific foraging associations. Ph.D. Thesis, University of California, Davis.
Strathman, R. R. 1978. The evolution and loss of feeding larval stages of marine invertebrates. Evolution 32: 894–906.
Suarez, S. S. 1975. The reproductive biology of *Ogilbia cayorum*, a viviparous brotulid fish. Bull. Mar. Sci. 25: 143–173.
Svärdson, G. 1949. Natural selection and egg number in fish. Rept. Inst. Freshw. Res., Drottningholm 29: 115–122.
Tavolga, W. N. 1954. Reproductive behavior in the gobiid fish *Bathygobius soporator*. Bull. Amer. Mus. Nat. Hist. 104: 431–459.
Thomson, J. M. & A. E. Bennett. 1953. The oyster blenny, *Osmobranchus anolius* (Valenciennes) (Blenniidae). Austr. J. Mar. Freshw. Res. 4: 227–233.
Thorson, G. 1946. Reproduction and development of Danish marine bottom invertebrates. Medd. Dan. Fisk. Havunders., Ser. Plankton 4: 1–523.

Thorson, G. 1950. Reproductive and larval ecology of marine botom invertebrates. Biol. Rev. 25: 1–45.
Tomlison, J. 1966. The advantage of hermaphroditism and parthenogenesis. J. Theoret. Biol. 11: 54–58.
Trivers, R. L. 1972. Parental investment and sexual selection. pp. 136–179. *In*: B. Campbell (ed.) Sexual Selection and the Descent of Man, 1871–1971. Aldine, Chicago.
Tyler, J. C. 1971. Habitat preferences of the fishes that dwell in shrub corals of the Great Barrier Reef. Proc. Acad. Nat. Sci. Philad. 123: 1–26.
Vitt, L.J. & J. D. Congdon. 1978. Body shape, reproductive effort, and relative clutch mass in lizards: resolution of a paradox. Amer. Natur. 112: 595–608.
Waddington, C. H. 1972. The evolution of an evolutionist. Edinburgh University Press, Edinburgh. 328 pp.
Ware, D. M. 1975. Relation between egg size, growth and natural mortality of larval fish. J. Fish. Res. Board Can. 32: 2503–2512.
Ware, D. M. 1977. Spawning time and egg size of Atlantic mackerel, *Scomber scombrus*, in relation to the plankton. J. Fish. Res. Board Can. 34: 2308–2315.
Warner, R. R. 1975. The adaptive significance of sequential hermaphroditism in animals. Amer. Natur. 109: 61–82.
Warner, R. R. 1978. The evolution of hermaphroditism and unisexuality in aquatic and terrestrial vertebrates. pp. 77–101. *In*: E. S. Reese & F. J. Lighter (ed.) Contrasts in Behavior, Wiley, New York.
Warner, R. R., D. R. Robertson & E. G. Leigh, Jr. 1975. Sex change and sexual selection. Science 19: 633–638.
Weatherly, A. H. 1972. Growth and ecology of fish populations. Academic Press, New York. 293 pp.
Williams, G. C. 1960. Dispersal of young marine fishes near Woods Hole, Massachusetts. Publ. Mus. Mich. St. Univ., Biol. Ser. 1: 329–368.
Williams, G. C. 1960. Adaptation and natural selection. Princeton University Press, Princeton. 307 pp.
Williams, G. C. 1967. Identification and seasonal size changes of eggs of the labrid fishes, *Tautogolabrus adspersus* and *Tautoga onitis*, of Long Island Sound. Copeia 1967: 452–453.
Williams, G. C. 1975. Sex and evolution. Princeton University Press, Princeton. 201 pp.
Wilson, E. O. 1975. Sociobiology: the new synthesis. Belknap, Cambridge, Massachusetts. 697 pp.
Wirtz, P. 1978. The behaviour of the Mediterranean *Trypterygion* species (Pisces, Blennioidei). Z. Tierpsychol. 48: 142–174.
Wourms, J. P. & O. Bayne. 1973. Development of the viviparous brotulid fish, *Dinematichthys iluceoteoides*. Copeia 1973: 32–40.
Wourms, J. P. & D. M. Cohen. 1975. Trophotaeniae, embryonic adaptations in the viviparous ophidioid fish, *Oligopus longhursti:* a study of museum specimens. J. Morphol. 147: 385–401.

Env. Biol. Fish. Vol. 6, No. 1, pp. 87-94, 1981

The reproductive behaviour of *Aequidens vittatus* (Pisces, Cichlidae) in Surinam, South America

Miles H. A. Keenleyside & Brian F. Bietz[1]

Department of Zoology, University of Western Ontario, London, Ontario N6A 5B7, Canada

Keywords:
Cichlids, Breeding behaviour, Parental care, Spawning site

Synopsis

The behaviour of free-living pairs of *Aequidens vittatus* was observed in Surinam, South America. An ethogram of breeding behaviour is presented, based on those observations and on aquarium-held fish. This is a biparental, substrate-guarding species that spawns on a movable platform. Sexual differentiation of parental roles was more pronounced when the young were in the embryonic interval than when they were mobile juveniles. Females spent more time near their brood, attacked juvenile conspecifics more often, and fanned and mouthed embryos more than males did. Males were more involved than females in territorial spacing of pairs. The movable spawning leaf provides support for eggs on detritus substrate and may also provide protection against predators and rapid water level changes.

Introduction

Reproductive strategies and male-female roles in parental care behaviour vary widely among fishes, and particularly in the family Cichlidae (Breder & Rosen 1966, Fryer & Iles 1972, Barlow 1974, Keenleyside 1979). Possible evolutionary explanations for this diversity have been discussed in several recent reviews (Ridley 1978, Blumer 1979, Perrone & Zaret 1979). Variation in cichlid breeding strategies appears to be especially pronounced in the primarily South American genera *Aequidens* and *Geophagus* (Keenleyside 1979). Many of these are biparental substrate-guarders, in which both parents collaborate in the care of their offspring close to the substrate for several weeks after spawning. At least one species (*Geophagus hondae*) is a maternal mouthbrooder; a pair spawns on the substrate and the female immediately picks up the eggs in her mouth and broods the young orally until they disperse (Sprenger 1971, personal observations). A few species combine substrate guarding and mouthbrooding; both parents guard the eggs on the bottom, but at hatching they pick up the young and brood them orally (e.g. *Aequidens paraguayensis* (Timms & Keenleyside 1975)).

There is also variation in the spawning substrate selected by *Aequidens* and *Geophagus* species. Many use a stationary site (e.g. a single stone, a gravel or sand patch, a leaf of a rooted plant), but some spawn on a loose leaf from the substrate litter. In these cases the parents, when disturbed, move their eggs by grasping the edge of the leaf in the mouth and swimming away with it. Use of a mobile

[1]Present address: MacLaren Marex, Inc., P.O. Box 9493, Station 'B', St. John's, Newfoundland A1A 2Y4, Canada

Received 3.3.1980 *Accepted 29.8.1980*

© Dr. W. Junk b.v. Publishers, The Hague. Printed in The Netherlands. 0378-1909/81/061-087 $01.60.

spawning site has been documented for *Aequidens coeruleopunctatus* (Barlow 1974), *A. paraguayensis* (Timms & Keenleyside 1975) and *Geophagus jurupari* (Cichocki 1976).

Detailed reports on reproductive behaviour of free-living South American cichlids are rare; they include those of Guimaraes (1930), Lowe-McConnell (1969) and Cichocki (1977). Little is known about the ecological conditions or the presumed selective forces associated with the variation in their reproductive strategies. One long-term goal of our research is to gather data in the field on the breeding behaviour and habitat of species with different breeding systems, and thereby to contribute to a better understanding of the evolution of cichlid reproductive behaviour.

In this paper we present data gathered during two trips to Surinam, South America. In April-May 1977, one of us (M.K.) made preliminary observations on *Aequidens vittatus* (Heckel), a biparental, substrate-guarding cichlid that uses a loose leaf as spawning substrate. In April-May 1979, we collected more extensive data on the same species. *A. vittatus* is widely distributed through north and central South America (Regan 1905, Eigenmann 1912, Eigenmann & Allen 1942, Lowe-McConnell 1969, Lüling 1971, Saul 1975). In Surinam it occurs in several of the major river systems, although its local, detailed distribution is not known (Boeseman personal communication).

Materials and methods

Description of study site

The study site is 24 km south of Paramaribo, the capital of Surinam, at approximately 5°37'N and 55°8'W. It is near the village of Onverdacht, at a former bauxite mine excavation, now filled with water and connected to the Para River, a tributary of the Surinam River. *A. vittatus* were found breeding in a shallow, weedy area close to shore, with no water current except for occasional surface movements caused by wind. The substrate was loose, flocculent, light greyish-brown detritus and silt, generally 6 to 8 cm deep, with firmer soil and gravel beneath. Aquatic vegetation consisted of dense to sparse stands of Cyperaceae, with a few Gramineae (*Paspalum* sp.) close to shore, and some patches of filamentous green algae. The approximate distribution of aquatic vegetation and water depth contours at the site are shown in Figure 1. The locations of guarding pairs while their behaviour was being monitored are also shown in Figure 1.

Data were collected at the beginning of the long rainy season (April and May). Occasional heavy rains caused rises of up to 13 cm in water depth at the site. Water temperatures ranged between 30 and 35° C. Several measurements of water quality, made with a LaMotte portable kit, are presented in Table 1. Data recorded in May, 1977, at two other locations in the Surinam River watershed where breeding *A. vittatus* were observed are included.

Other fishes present at the main study site were *Cichlasoma bimaculatum* (several breeding pairs), *Crenicichla* sp. (two size groups: 6 - 8 cm and 10 - 15 cm), *Hoplias* sp., and several characins, including *Copella arnoldi* and *Hemigrammus armstrongi*.

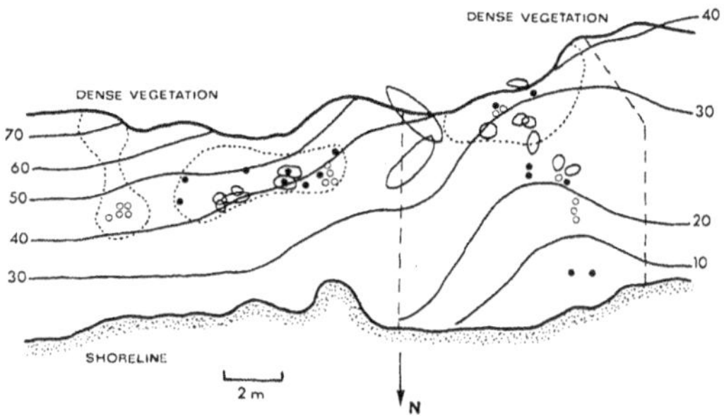

Fig. 1. Scale drawing of study site showing locations of *A. vittatus* pairs guarding pre- and post-hatching embryos or free-swimming young (solid circles, open circles and larger enclosed areas respectively). Three zones enclosed by dotted lines had no rooted aquatic plants. Remainder of site had many rooted aquatics, with relative density lowest in area between the two broken lines. Depth contours (in cm) were recorded on 13 May, 1979.

Data collection

Data were collected as follows: each day in the early morning one of us walked slowly through the site until a pair of *A. vittatus* with young was found. The observer remained still for 5 min to allow the fish to recover from disturbance. Then one parent was watched closely for 10 min, during which selected behaviours (see below) were monitored and called out to the second observer who recorded data in a notebook. The other parent was then watched

Table 1. Water analysis at Para River study site and two nearby locations.

Location	Date	Dissolved O_2 (ppm)	pH	Dissolved solids (ppmCaCO$_3$)	Total hardness (ppmCA + MG)	Turbidity (Jackson units)
Para River	21/04/79	5.6	6.0–6.5	190	< 10	25
Para River	25/04/79	5.0	5.5	150	6	10
Para River	30/04/79	5.8	6.0–6.5	90	6	5–10
Para River	01/05/79	5.0	6.0–6.5	100	4	5
Para River	05/05/79	5.8	6.0–6.5	150	6	< 5
Cola Creek	30/04/77	–	4.5–5.0	190	< 2	< 5
Km 43 Pond	08/05/77	–	7.0	110	15	45

for 10 min and its behaviour recorded in the same way. The location of the family unit was recorded on a study site map. Data were collected in this manner from 15 pairs with clutches at each of the pre- and post-hatch non-swimming young, and free-swimming juvenile intervals, for a total of 900 min recording time.

Behaviours recorded were: time spent within one body length of the young; frequency and duration of fanning bouts (non-swimming young only); frequency of mouthing the young, moving the spawning leaf, feeding acts, and attacks directed at other fish. Also recorded were the identity of attacked fish and the approximate distance from the young to the aggressive interaction.

Adult *A. vittatus* captured at the study site were taken to our laboratory in London, Ontario. Some pairs were observed breeding in aquaria, allowing more detailed observations of courtship, spawning, and the interactions between young and their parents.

Results

Description of the species

Aequidens vittatus is a medium-sized cichlid. Mean total length of 23 breeding-size fish collected at the study site was 10.1 cm (S.D. = 0.76; range = 8.8 - 11.2). Within breeding pairs the male was usually slightly larger than the female and his dorsal and anal fin rays were more extended. Fish from different locations varied somewhat in general colouring, probably reflecting local differences in water quality and diet. Those at the study site were light olive-brown dorsally, shading to silver ventrally.

Non-breeding fish have a dark stripe on each side extending from the eye forward to the snout and backward to the base of the posterior dorsal fin rays, a faint dark bar extending downward from the eye to the preopercular angle, a dark spot at the base of the dorsal caudal fin rays, and a large dark patch below the 7th to 9th dorsal fin spines and overlapping the lateral stripe. The fins are pale yellow to transparent. There are light blue spots on the median fins, and the most anterior pelvic fin rays are bright yellow.

Breeding adults differ in several respects. The major change is caused by fading of the lateral stripe and the appearance of five or six vertical dark bars, the most anterior passing through the large, mid-lateral patch. These bars are separated by narrow, pale yellow bars. The mid-lateral patch and the spot at the caudal base become intensely black. The eye bar also becomes darker, and extends dorsally through the eye. The speckling on the median fins becomes more intense blue, and a light turquoise line overlies the yellow anterior pelvic rays. The dorsal fin of males has reddish-orange edging.

Ethogram of breeding behaviour

a) Pair formation and spawning site selection

The first indication that *A. vittatus* were preparing to spawn was the persistent close association of two fish, moving slowly about near the bottom in relatively open water. They were often in moderately bright breeding colours, which intensified when they interacted with other fish. Usually the pair remained close to a leaf lying on the substrate. The female occasionally picked at it or grasped the edge in her mouth and moved it a short distance. They were aggressive towards approaching conspecifics, keeping them away from the immediate vicinity of the leaf.

Typically such a pair moved about, associating with several leaves in turn. Eventually they remained with one leaf and began to show more intense territorial behaviour around it. Either one fish (usually the male) or both moved out to challenge nearby conspecifics and chase them away. Such encounters were often followed by brief bouts of leaf-tugging by one or both fish. By this stage it was clear that a pair had formed and had selected a spawning site.

b) Courtship and spawning

Interactions between the members of the pair gradually increased in frequency and were often centred over the leaf. Typically the male moved into a broadside position in front of the female, fully spread all his fins, and vigorously directed several tail beats at her. She responded by spreading all her fins and tail-beating at the male. This was often followed by a brief, vigorous bout of mutual tight circling.

Bouts of body quivering then began. One fish pitched slightly head-down and the entire head and body quivered, sometimes for as long as 4 or 5 seconds. Quivering bouts were usually performed over the leaf, and as they increased in frequency the quivering fish often contacted the leaf with its mouth, or picked up the leaf and held it off the bottom while quivering. This conspicuous activity was clearly directed at the other member of the pair, and was most often done by the female. She occasionally picked up the leaf and maneuvered to keep it in front of the male while she quivered.

By this stage the genital papilla of both fish was clearly protruding. The female, and less often the male, began cleaning the leaf by nipping at its surface. Shortly before spawning she performed bouts of skimming over the leaf, brushing it with her large, blunt, genital papilla. Between bouts of skimming and leaf cleaning, she courted with the male, both fish quivering above the leaf, and the male occasionally tail-beating towards her.

When spawning began the female made one or more passes over the leaf, releasing a string of eggs on each pass, then moved aside as the male swam slowly over the leaf, presumably releasing sperm. This continued until the clutch was laid.

c) Parental care

The parental behaviour of *A. vittatus* pairs is similar to that of other biparental substrate-guarding cichlids. They aerated and removed fine particles from the eggs by holding position slightly to one side of the egg-bearing leaf, and fanning vigorously with the pectoral fins. Occasionally they gently picked at the eggs with the mouth, thus removing foreign particles. Dead eggs were removed by more vigorous mouth-picking. While fanning, females were usually 1 to 2 cm, males 5 to 7 cm above the eggs. When disturbed the female (very seldom the male) moved the clutch by tugging the leaf away from the source of disturbance. Occasionally both parents stayed close to the eggs; more often they exchanged roles, one staying with the eggs while the other swam nearby, chasing away intruders, feeding, or cruising slowly about.

Before the embryos hatched the parents began excavating one ore more small pits near the clutch, by using the mouth to carry away substrate particles. At hatching the young were carried orally by both parents to one of the pits, where they were fanned and mouthed in much the same way as before hatching. The young were occasionally moved from one pit to another.

As the young developed into free-swimming juveniles the family unit became more mobile. The parents were very attentive, both staying close to the brood most of the time. Often they faced away from each other, tail to tail among the school of young, as though maximizing the extent to which they could visually scan the surroundings. When alarmed the parents signalled their young by rapidly extending forward and retracting their pelvic fins and by jerking laterally the head and body in a quick, twitching movement. The young responded by dropping to the substrate and clustering around the signalling adult. Pairs with eggs or hatched embryos were found only in the small, plant-free areas, or at the western end of the site, where aquatic vegetation was sparse (Fig. 1). Those with mobile young were more widely distributed; the larger the young the more likely the family unit was to be in dense vegetation. Mobile young were not seen without adults; apparently as they became independent they left the study site or remained hidden among plants.

Parental care data

Data on three aspects of the parental care behaviour of male and female *A. vittatus* during three developmental intervals of their young are presented in Table 2. The behaviours were: 1) time spent close to the brood, 2) frequency of attacks at three classes of intruding fish (conspecific adults, conspecific juveniles, and fish of other species), and 3) approximate straight-line distance between the brood and the point of interaction with intruders. A 2 × 3 factorial MANOVA was used initially to test for overall differences in behaviour between sexes and among developmental intervals. MANOVA was chosen since repeated measures were made on some fish and since the variables measured were not independent. The presence of a highly significant interaction term prevented main effect comparisons within the MANOVA framework. Therefore, a univariate 2-way ANOVA was then used, despite the potential for inflated error terms as the result of non-independence of variables (Winer 1971). All variables except time were log-transformed before analysis, in order to approximate homogeneity of variances.

Significant interaction between sex and developmental interval was found for two variables: time with young, and distance travelled to attack adult *A. vittatus* (Table 2). While their progeny were at the pre- and post-hatching embryonic intervals, male parents spent less time with the clutch, and swam further away to attack adult conspecifics than females did. There was much less difference between the sexes in these two measures when they were guarding mobile young. Only one other sex-related difference in parental behaviour was observed; females attacked conspecific juveniles at a higher rate than males did, especially during the post-hatching embryonic interval.

The overall frequencies of attacks at adult conspecifics did not show consistent male-female differences (Table 2). However, when the frequencies of such attacks were partitioned according to the colour pattern of the attacked fish, a sex-related difference did appear (Table 3). Males directed a higher proportion of their attacks at adult conspecifics in breeding colouration than females did, especially when broods were in the pre- and post-hatching embryonic intervals (chi-square = 8.5; $p < 0.05$). In general, however, both male and female

Table 2. Time spent close to the clutch, and aggressive behaviour of parental *A. vittatus*. Values are means (1 SE) per 10-min recording period. n = 15 for each developmental interval.

Behaviour	Sex	Eggs	Post-hatching embryos	Free-swimming young
Time with clutch,	♀	7.8 (0.33)	7.4 (0.44)	8.1 (0.36)
min*	♂	2.5 (0.44)	3.3 (0.43)	8.1 (0.37)
Frequency of	♀	2.9 (0.62)	3.3 (1.00)	3.3 (0.94)
attacks at adults	♂	3.3 (0.70)	3.7 (0.55)	2.2 (0.59)
Freq. of attacks	♀	1.8 (0.53)	3.8 (0.71)	2.0 (0.68)
at juveniles**	♂	1.0 (0.24)	0.9 (0.31)	1.3 (0.48)
Freq of attacks at	♀	0.3 (0.16)	0.5 (0.17)	0.6 (0.21)
other species	♂	0.3 (0.15)	0.2 (0.14)	0.4 (0.16)
Distance to	♀	38.0 (6.88)	32.6 (7.53)	34.5 (5.46)
attack adults, cm*	♂	77.1 (11.15)	70.2 (7.57)	21.1 (4.83)
Distance to	♀	29.3 (8.25)	39.6 (6.36)	28.9 (5.85)
attack juveniles	♂	29.6 (6.46)	29.6 (8.28)	16.7 (5.35)
Distance to attack	♀	7.7 (3.61)	15.0 (6.56)	21.0 (8.44)
other species	♂	8.8 (5.01)	7.8 (5.41)	12.7 (6.43)

* Highly significant interaction between sex and developmental interval ($p < 0.001$).
** Highly significant difference between sexes ($p = 0.001$).

Table 3. Total frequency and percentage (in parentheses) of attacks by guarding *A. vittatus* directed at adult conspecifics in breeding (B) and non-breeding (NB) colouration, during the three developmental intervals of their broods.

Parent	Developmental interval					
	Eggs		Post-hatching embryos		Free-swimming young	
	B	NB	B	NB	B	NB
♂	14 (28.0)	36 (72.0)	10 (17.9)	46 (82.1)	14 (45.2)	17 (54.8)
♀	5 (11.6)	38 (88.4)	1 (2.0)	49 (98.0)	17 (34.7)	32 (65.3)

parents attacked other adults in non-breeding colours more often than those in breeding colours.

Potential clutch predators of other species (primarily *Crenicichla* and *Hoplias*) were not attacked unless they swam close to the clutch. Adult *A. vittatus* appeared to ignore the numerous small characins in the area.

Several other behaviours performed by guarding adults are summarized in Table 4. Three separate measures of fanning activity show the same male-female difference. Whether measured as total fanning time, number of fanning bouts per observation period, or mean duration of fanning bouts, females fanned more than males. This difference was most pronounced when the progeny were in the egg interval, but persisted during the post-hatching embryonic interval, except for mean bout duration. Mouthing of embryos was almost entirely a female activity; only two males showed any mouthing. Only females were recorded as moving leaves with eggs, and this activity was not seen frequently. Males performed more feeding acts than females during all three developmental intervals, but this difference was significant only during the post-hatching embryonic interval.

Discussion

The general pattern of reproduction in *A. vittatus* is typical of substrate-guarding, biparental New World cichlids. Brood-rearing is an actively co-operative, joint enterprise, with partial division of labour between the members of the pair, especially before the progeny become free-swimming. Female parents spent more time close to the clutch, and they performed more fanning, mouthing and leaf-moving than males did, as has been reported for other species (e.g. Barlow 1974, Timms & Keenleyside 1975, Smith-Grayton & Keenleyside 1978).

In general, males of substrate-guarding cichlid pairs are believed to be more active than females in aggressive defence of their young (Barlow 1974, Keenleyside 1979). This has been confirmed for breeding pairs of *Cichlasoma maculicauda* in Gatun Lake, Panama, but only while adults were guarding eggs; among pairs with free-swimming young, female parents attacked intruders more often than males did (Perrone 1978). Our data show an even more complex relationship. For example, female parents consistently attacked juvenile *A. vittatus*

Table 4. Various behaviours of parental *A. vittatus*. Values are means (ISE) per 10-min recording period. $n = 15$ for each developmental interval.

Behaviour	Sex	Eggs	Post-hatching embryos	Free-swimming young
Fanning time, sec	♀	284.5 (27.47)	146.9 (36.16)	0
	♂	61.8 (9.28)***	52.5 (16.56)*	0
Fanning bout frequency	♀	6.9 (0.71)	4.6 (0.95)	–
	♂	2.7 (0.32)***	1.9 (0.57)*	–
Bout length, sec	♀	51.3 (12.70)	25.3 (5.04)	–
	♂	21.2 (2.38)*	15.9 (5.33)	–
Freq. of mouthing young	♀	4.4 (1.94)	4.0 (1.25)	0
	♂	0.1 (0.09)	0.3 (0.27)	0
Freq. of moving leaf	♀	0.7 (0.25)	–	–
	♂	0	–	–
Feeding freq.	♀	2.2 (0.57)	1.9 (0.49)	2.7 (0.56)
	♂	4.1 (0.83)	3.7 (0.71)*	3.0 (0.70)

* $p < 0.05$ (paired sample t-test).
*** $p < 0.001$.

more often than males did, whereas both parents attacked adult conspecifics and fish of other species with about equal frequency. However, males moved further from their clutch than females did to challenge adult conspecifics, and directed a higher proportion of these attacks at adults in breeding colouration. Therefore, some male aggression appeared to be related to territorial spacing of breeding pairs, rather than to repulsing potential predators. Males spent much time cruising at some distance from their brood, occasionally interacting with other breeding adults, but apparently ignoring juveniles. Only when the latter approached a brood closely were they attacked by the attending parents, and then usually by the female.

The different frequencies with which breeding and non-breeding fish were attacked by guarding parents (Table 3) is explained by the greater mobility of the nonbreeders. Brood-guarding fish were more site-attached and therefore less likely to enter the territory of other guarding pairs than were the nonbreeders who moved freely throughout the study site.

Quantitative differences in behaviour between the parents declined when the progeny became free swimming. Both parents then stayed close to the young, moving slowly as the school of foraging young moved, and apparently scanning the surroundings for potential predators. The need for both adults to guard their mobile young was emphasized on two occasions when we removed one member of a guarding pair. Conspecific adults and juveniles quickly gathered at the site, where they soon ate the young despite vigorous aggression by the remaining parent.

Non-breeding *A. vittatus* were often seen feeding at the study site. By comparison, the feeding frequency of guarding adults was low (Table 4), even though food was presumably available throughout the site, including the immediate vicinity of the clutches. The tendency of males to feed more often than females, especially before their young were mobile, was probably related to the differential fidelity of the parents to the clutch site.

The use of a portable spawning platform is typical of reproduction in *A. vittatus* and is not limited to our study site. In 1977 a pair was seen guarding eggs on a loose leaf in a quiet backwater of a fast-flowing forest creek. In that location the eggs would seem to have been more secure on a stone or one of the numerous tree roots protruding from the stream bank. However, protection against predation and rapid water level changes (which often occur in *A. vittatus* habitat in Surinam following torrential rains) might be enhanced by use of a mobile spawning site. Both of these functions have been suggested for other litter-spawning *Aequidens* species (Lowe-McConnell 1964, Barlow 1974, Keenleyside & Prince 1976). *A. vittatus* adults will quickly move their egg-bearing leaf away from a source of disturbance, but we have no evidence that sudden changes in water depth induce leaf movements. At our Para River study site, scattered leaves appeared to provide the only firm support for a clutch of eggs on the detritus-covered bottom. Their presence may have induced the fish to spawn there, since breeding *A. vittatus* were not found further from shoreline vegetation, where the major habitat difference seemed to be the lack of leaf litter on the substrate.

Adults guarding embryos occurred only in areas relatively free of vegetation (Fig. 1), suggesting that a clear view of at least part of the immediate surroundings is an important aspect of suitable clutch-rearing microhabitat, presumably because predators can be detected at a distance from the clutch. Once the young have become free-swimming they are probably less vulnerable to predation, since they can dive for cover in the substrate. We occasionally observed adults with mobile young in denser vegetation (Fig. 1), but larger, independent young were never seen in open areas.

We are encouraged by these results to continue the study of breeding behaviour of free-living cichlids in South America. In particular we plan to study mouth-brooding and combination-brooding species (those that guard their eggs on the substrate but carry their post-hatch young orally) in relation to their breeding habitats, to gain a better understanding of the evolution of these different parental strategies.

Acknowledgements

We are extremely grateful to all those people who helped us with our work in Surinam. In particular we thank: Mr. F.F. van Dijk, Head of the Fisheries

Division, Ministry of Agriculture, Animal Husbandry and Fisheries; Mr. J. Douglas, Director of the Stichting Surinaams Museum; and Mr. H. Heyde of Paramaribo. Dr. J. B. Phipps and Dr. T. Dickinson kindly helped with plant identification. James Grant was an indispensable assistant on the 1977 survey trip. We thank Drs. J.-G. Godin and B. H. Seghers for helpful comments on the manuscript. The work was supported by operating and travel grants from the Natural Sciences and Engineering Research Council of Canada.

References cited

Barlow, G. W. 1974. Contrasts in social behavior between Central American cichlid fishes and coral-reef surgeon fishes. Amer. Zool. 14: 9–34.

Blumer, L. S. 1979. Male parental care in the bony fishes. Quart. Rev. Biol. 54: 149–161.

Breder, C. M. & D. E. Rosen. 1966. Modes of reproduction in fishes. Nat. Hist. Press, Garden City. 941 pp.

Cichocki, F. 1976. Cladistic history of cichlid fishes and reproductive strategies of the American genera *Acarichthys*, *Biotodoma* and *Geophagus*. Ph.D. thesis, University of Michigan, Ann Arbor. 710 pp.

Cichocki, F. 1977. Tidal cycling and parental behavior of the cichlid fish, *Biotodoma cupido*. Env. Biol. Fish. 1: 159–169.

Eigenmann, C. H. 1912. The freshwater fishes of British Guiana, including a study of the ecological grouping of species and the relation of the fauna of the plateau to that of the lowlands. Mem. Carneg. Mus. 5: 1–578.

Eigenmann, C. H. & W. R. Allen. 1942. Fishes of Western South America. Univ. Kentucky Press, Lexington. 494 pp.

Fryer, G. & T. D. Iles. 1972. The cichlid fishes of the Great Lakes of Africa. T. F. H. Publishers, Neptune City. 641 pp.

Guimaraes, J. R. A. 1930. O Acara. *Geophagus brasiliensis*, Quoy et Gaimard. Rev. Indust. Anim. 1: 658–662.

Keenleyside, M. H. A. 1979. Diversity and adaptation in fish behaviour. Zoophysiology, Vol. 11. Springer-Verlag, Heidelberg. 208 pp.

Keenleyside, M. H. A. & C. E. Prince. 1976. Spawning-site selection in relation to parental care of eggs in *Aequidens paraguayensis* (Pisces: Cichlidae). Can. J. Zool., 54: 2135–2139.

Lowe-McConnell, R. H. 1964. The fishes of the Rupununi savanna district of British Guiana, South America. Part I. Ecological groupings of fish species and effects of the seasonal cycle on the fish. J. Linn. Soc. Lond. Zool. 45: 103–144.

Lowe-McConnell, R. H. 1969. The cichlid fishes of Guyana, South America, with notes on their ecology and breeding behaviour. Zool. J. Linn. Soc. 48: 255–302.

Lüling, K. H. 1971. *Aequidens vittatus* (Heckel) und andere Fische des Rio Huallaga im Übergangsbereich zur Hylaea. Zool. Beitr. 17: 193–226.

Perrone, M. 1978. Mate size and breeding success in a monogamous cichlid fish. Env. Biol. Fish. 3: 193–201.

Perrone, M. & T. M. Zaret. 1979. Parental care patterns of fishes. Amer. Nat. 113: 351–361.

Regan, C. T. 1905. A revision of the fishes of the South-American cichlid genera *Acara*, *Nannacara*, *Acaropsis*, and *Astronotus*. Ann. Mag. Nat. Hist. (ser. 7), 15: 329–347.

Ridley, M. 1978. Paternal care. Anim. Behav. 26: 904–932.

Saul, W. G. 1975. An ecological study of fishes at a site in upper Amazonian Ecuador. Proc. Acad. Nat. Sci., Philad. 127: 93–134.

Smith-Grayton, P. K. & M. H. A. Keenleyside. 1978. Male-female parental roles in *Herotilapia multispinosa* (Pisces: Cichlidae). Anim. Behav. 26: 520–526.

Sprenger, K. 1971. The red hump *Geophagus*. Buntb. Bull. 26: 24–25.

Timms, A. M. & M. H. A. Keenleyside. 1975. The reproductive behaviour of *Aequidens paraguayensis* (Pisces, Cichlidae). Z. Tierpsychol. 39: 8–23.

Winer, B. J. 1971. Statistical principles in experimental design. 2nd ed. McGraw-Hill, San Francisco. 907 pp.

Env. Biol. Fish. Vol. 6, No. 1, pp 95–103, 1981

Reproductive tactics of the Asian cichlids of the genus *Etroplus* in Sri Lanka

Jack A. Ward & Jayampathy I. Samarakoon

Department of Biological Sciences, Illinois State University, Normal, Illinois 61761, U.S.A.

Keywords:
Behavior, Ecology, Parental care, Reproductive cycles, Colonial nesting, *Etroplus maculatus*, *Etroplus suratensis*

Synopsis

The Asian cichlids, *Etroplus maculatus* (the orange chromide) and *E. suratensis* (the green chromide) in Sri Lanka reproduce twice during the year when water conditions are favorable for nest construction and maintaining visual contact with offspring. These are the drier premonsoonal and monsoonal seasons when water turbidity decreased and salinity increased. When breeding in isolation orange chromide pairs selected dense vegetation where nests were camouflaged. During the peak breeding cycle (July) orange chromide pairs selected sparse vegetation for nesting as a compromise between survival of young and availability of adult food. These areas were also occupied by foraging non-breeding conspecifics which increased the threat of cannibalism of offspring. Under these pressures most orange chromides nested in colonies which helped decrease both actual and attempted cannibalism. The sympatric green chromide does not forage during nesting and nest site selection was determined mainly by factors favoring offspring survival. Biparental care is exhibited by both species. One member of an orange chromide pair stands guard over offspring while the other leaves the territory to forage -- their roles are reversed every few minutes. The total parental investment is equivalent to the full investment of a single parent. Monogamy appears to be maintained by their metabolic constraints. In the green chromide both parents are vigilant over offspring and neither forages thus spending twice as much time in parental investment.

Introduction

Other contributions to this symposium (Barlow 1981, Borowsky & Diffley 1981, Keenleyside & Bietz 1981, Dominey 1981) have described a variety of reproductive tactics among species ranging from bluegills to coral reef fishes, as have several recent reviews (Blumer 1979, Ridley 1978).

The relationships between the energy expended in acquiring mates and the individual investments by males and females in caring for offspring, have been suggested by some as the cornerstones of reproductive success (Trivers 1972, Williams 1975). Perrone & Zaret (1979) concluded that biparental care among some fishes (e.g., cichlids) is a serious evolutionary problem because in terms of investment both parents are not required to raise a brood. We report here our field studies of biparental care displayed by the cichlids *Etroplus maculatus* (the orange chromide) and *E. suratensis* (the green

Received 20.6.1980 *Accepted 28.8.1980*

© Dr. W. Junk b.v. Publishers, The Hague. Printed in The Netherlands. 0378-1909/81/061-095 $01.80

chromide) with particular regard to time of reproduction, breeding habitat, nest distribution, nest location, behavioral time budgets, territoriality and predation of young. When these factors are considered simultaneously, as they occur in nature, then the biparental care exhibited by these species is not a serious evolutionary problem but an adaptation necessary for survival. *E. maculatus* and *E. suratensis*, the only indigenous cichlids of Asia, occur in southern India and the island of Sri Lanka (Ceylon). We have studied them only in Sri Lanka where they are sympatric in both brackish water coastal estuaries and freshwater inland lakes. Both species have been previously studied in various laboratory investigations and some field studies (Ward & Wyman 1975, 1977).

Materials and methods

This field study began in August 1976 and ended in January 1978, but was not (and could not be) continuous because of weather constraints. However, a 1 hectare study area was monitored for a full 12 months in shallow water along the western shore of Negombo Lagoon on the southwestern coast of Sri Lanka (Fig. 1), in a bar-built estuary (Pritchard 1967). Major variation in salinity is horizontal but stratification ranging from 0 – 8 ppt depending on season and location is possible (DeSilva & Silva 1979). Two rainy seasons extended from the middle of April to late May and again from late September until November. The latter period is the convectional cyclonic rains that follow southwest monsoons which in Sri Lanka are not accompanied by heavy rainfall (Peiris 1977). Two relatively dry seasons extend from December to the middle of April (the permonsoonal season) and from June to September (the monsoonal season) and alternate with the rainy seasons.

The most prominent features of the study area were three brush piles, man-made fishing traps made of mangrove branches where fish take refuge (Ward & Wymann 1975, 1977). These areas were the foci of much of the nesting activities to be reported.

All observations (more than 1300 hours total) were made underwater using a face mask and a snorkel. Fish nests were in shallow water from 30–110 cm deep. Field notes and quantitative behavioral data were written on water resistant paper attached to a plastic sheet. Some behavior was recorded permanently on 35 mm and super-8 motion picture film using a Pentax ES II camera and Canon Auto Zoom 514 super-8 camera in Ikelite housings (Ikelite, Indianapolis).

During the study period hydrographic and meteorological data were collected regularly and the distribution of vegetation in the study area was mapped (to be reported elsewhere). The entire study area was surveyed every third day while swimming along transects at 10 m intervals. We are sure that most isolated nests were located by this method and that colonial nesters never escaped detection. Nests in sparse vegetation were marked by dropping 10 mm metal nuts with colored plastic ribbons near the location of each nest. Nests in dense vegetation or among mangrove litter were marked by planting sticks nearby that projected above the surface of the water. The locations of all nests were also logged on a map as soon as possible after location.

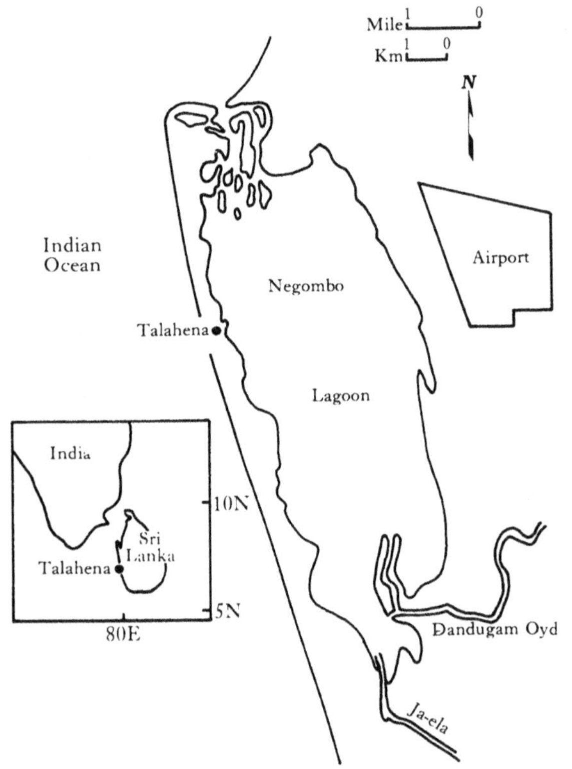

Fig. 1. Map of Negombo Lagoon. The hectare study area is situated 18 km east of Blue Lagoon Hotel, Talahena. The inset shows geographic location of Sri Lanka.

Distances of nearest neighbors to nesting pairs were measured at all nests using a two meter rule. Nearest neighbor distance was dynamic, particularly during the eleutheroembryonic phase of ontogeny (Balon 1975) when parents frequently transferred embryos to newly dug pits. Home range of nesting pairs was also estimated. However, home ranges could only be mapped for pairs in shallow water and then only when surface waters were not disturbed by wind. Home range distance was approximated by standing and looking in the water from a position where two pairs of nesting fish could be observed at once. One pair was observed for 15 min and then at 3 min intervals a marker was dropped at the farthest position one of the partners wandered from the center of the nest during that period of time. The marker was dropped in position by a long pole; the markers were 10 mm metal nuts with loops of colored ribbon, a different color for each pair.

Reproductive success was also measured for nesting pairs. In this study it was defined as pairs that protected eggs and the subsequent eleutheroembryos until they became free-swimming young. After that parents and young often became nomadic and abandoned the nest site making it impossible to determine their fate. Unsuccessful pairs lost eggs or eleutheroembryos to cannibalism or other disturbance and subsequently abandoned their nest sites.

Time budgeted by pairs were determined for four activities: tending the offspring (fanning or mouthing eggs and eleutheroembryos), holding station (hovering over eggs and eleutheroembryos), extra-pair interactions (interactions with conspecific and heterospecific intruders), and wandering and foraging. Time budget data were collected only between 1000 and 1400 h when underwater visibility was best. Each category of behavior was timed during a 2 min period for each sex – then the next category was timed and so forth. A stopwatch in an underwater housing was used. Time budgets were constructed from data collected for 10 pairs.

Results

Time of reproduction

Both *E. maculatus* and *E. suratensis* nested and reproduced during both the premonsoonal and monsoonal seasons. Since they are substrate guarders counting nests containing eggs or eleutheroembryos was a reliable index of active reproduction. *E. maculatus* bred primarily during February and July (Fig. 2) while *E. suratensis* bred twice during the premonsoolal season, during January and April, and again in July (Fig. 2). Breeding frequency for both species increased threefold during July (Fig. 2).

Reproduction occurred when water conditions were favorable for nest construction and maintaining visual contact between parents and offspring. The water was murky during the rainy seasons but progressively cleared until underwater visibility during breeding peaks exceeded 3 meters. Heavy river discharge during rainy seasons caused strong seaward currents which dislodged some vegetation in the shallow areas. During breeding peaks only tidal currents prevailed but they did not noticeably disturb the substratum. As river discharge diminished tidal exchange caused salinity to rise from 5 ppt to 18–30 ppt during breeding peaks.

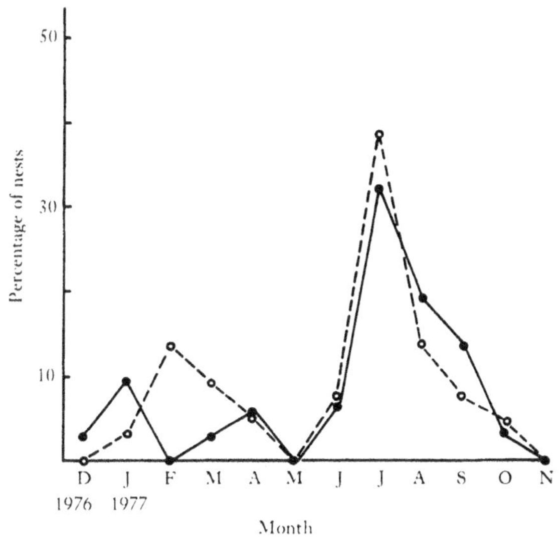

Fig. 2. Monthly relative frequency distribution of nests for *E. maculatus* (○) and *E. suratensis* (●).

E. maculatus: Breeding habitat and nest distribution

Orange chromides bred in several patches that differed in amount of cover, spawning surface available, water depth and so forth. Cover, or rooted vegetation, was designated as sparse (substratum visible between plants) or dense (substratum obscured by prolific growth of plants). During peak periods of breeding in both seasons both microhabitats were available and often as contiguous patches. During the peak breeding of the premonsoonal season, 21 February to 7 March, more nests (68.18%) were constructed in sparse patches (Fig. 3, Table 1), however, this difference was not statistically significant. More pairs, 66.65%, nesting in sparse patches were reproductively successful than pairs (42.86%) in dense vegetation but again this difference was not statistically significant. After the breeding peak significantly more pairs nested in dense vegetation (Table 1, $X^2 = 6.3$, $p \leqq 0.05$). Because of fishing activity it was impossible to determine if these pairs were reproductively successful.

During the peak breeding of the monsoonal season, 26 June to 4 August, a total of 60 nests were constructed in the vicinity of brush piles II and III. Significantly more nests were constructed around III (Table 2, $X^2 = 5.40$, $p \leqq 0.05$). It was here that extensive observations were made on 39 pairs. Significantly more of these nests were constructed in sparse patches (Table 2, $X^2 = 18.22$, $p \leqq 0.05$); occupants of these nests were reproductively more successful ($X^2 > 4.18$, $p \leqq 0.05$). During 4 August to 13 October, after the breeding peak, had passed 30 nests were traced but they were indiscriminately distributed among sparse and dense patches, but now reproductive success was greater for pairs nesting in dense vegetation (Table 2, $X^2 = 5.79$, $p \leqq 0.05$).

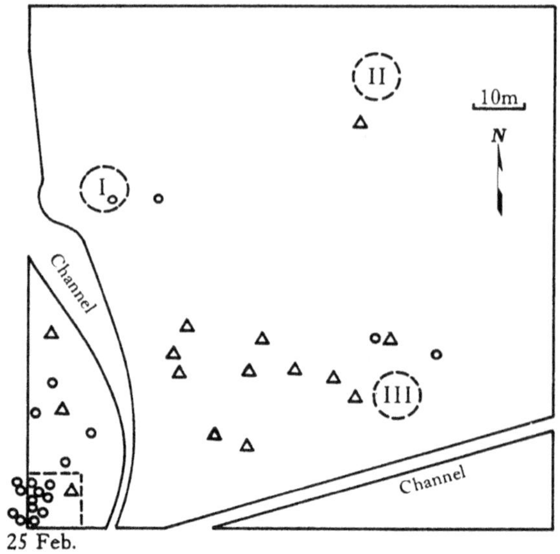

Fig. 3. E. maculatus nest dispersion during peak (o) and after the peak (△) for premonsoonal dry season. The cluster of 12 nests (○) at the southwestern corner is the colony as it appeared on 25 February 1977. Roman numerals encircled by broken lines are brushpiles.

Some of the nests described above were isolated while others were aggregated into colonies. Below we compare reproductive success for colonial nesters vs. nests in isolation.

During the premonsoonal breeding peak 15 nests were aggregated into a colony. The distance to the nearest neighbor in the colony was 1.83 m ($n = 8$, $SE = 0.7$). For 7 isolated nests nearest neighbor

Table 1. Nests and nest distribution of *E. maculatus* during the premonsoonal dry season.

Nesting period	Vegetation distribution	Nest dispersion	Number of nests	Number of spawns lost as eggs	Number of spawns lost as eleutheroembryos	Number of spawns reared to free swimming
Nesting peak 21 Feb. to 7 March	Sparse	Colony	15	1	4	10
	Dense	Isolated	7	2	2	3
24 March to 21 Arpil	Sparse	Isolated	3*	—	—	—
	Dense	Isolated	12*	—	—	—

* The fate of these nests could not be determined because of interference by fishermen.

Table 2. Nests and nest distribution for *E. maculatus* during the monsoonal dry season.

Nesting period	Vegetation distribution	Nest dispersion	Number of nests	Number of spawns lost as eggs	Number of spawns lost as eleutheroembryos	Number of spawns reared to free swimming
Nesting peak 26 June to 4 August BP III *	Sparse	Colony	34	8	3	23
	Dense	Colony	5	4	0	1
26 June to 4 August BP II	Sparse & Dense	Colony	21	6	4	11
4 August to 13 October BP III	Sparse	Isolated	12	7	1	4
4 August to 13 October BP II	Sparse	Isolated	6	3	0	3
4 August to 13 October BP III & Open **	Dense	Isolated	12	0	2	10

* BP = brush pile; ** open = >5 m from periphery of brush pile.

distance exceeded 5 m. More pairs in the colony were reproductively successful but the difference was not significant. A more precise comparison became possible during the monsoonal season. During peak breeding of the monsoonal season three colonies were apparent near brush pile III. Distances to nearest neighbors were 1.78 m (n = 9; SE = 0.14), 1.82 m (n = 6; SE = 0.11) and 2.05 m (n = 13; SE = 0.14). After the peak of breeding 12 isolated nests were constructed in the same patches. Reproductive success of the colonial nesters was significantly higher than that for isolated nesters (Table 2, $X^2 = 4.30$; $p \leqq 0.05$). Although the structural habitat remained the same during colonial and isolated nesting, the biotic environment changed; conspecific abundance was conspicuously greater during colonial nesting and therefore the threat of cannibalism proportionately higher.

E. suratensis: Breeding habitat and nest distribution

The pattern of nest site selection for the green chromide was influenced by vegetation, availability of spawning sites and the distribution of orange chromides. Green chromide nests were always isolated.

During the premonsoonal season breeding began in December 1976. The first nesting pair was located in a clearing near brush pile II. In the following months six other breeding pairs were observed. Five nests were in the northwestern sector (Fig. 4). There vegetation consisted almost ex-

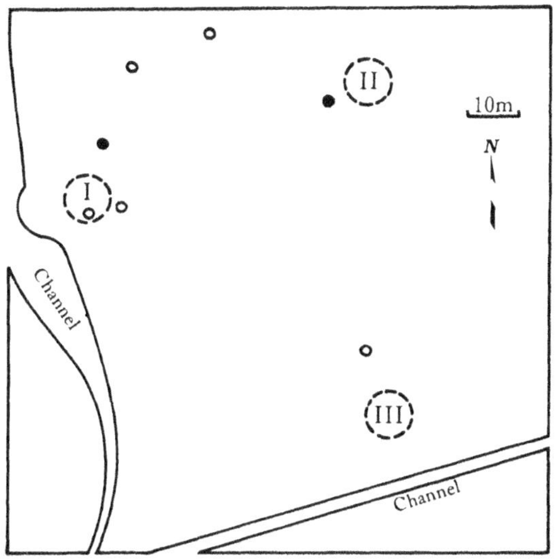

Fig. 4. E. suratensis nest dispersion during premonsoonal dry season showing nests of successful (●) and unsuccessful (○) pairs. Roman numerals encircled by broken lines are brushpiles.

clusively of the short sea grasses, *Halophila* spp., however, the nests were located within bare areas about one meter in diameter fringed with vegetation. Each clearing had a prominent object for egg attachment which was either a coconut husk or a mangrove stem. The area selected for nesting had very few nesting orange chromides.

During the monsoonal season two contrasting types of habitat were available. One had abundant rooted vegetation but also supported a large population of orange chromides. The green chromide began breeding in June. A total of 24 breeding pairs were observed. Eleven constructed their nests in the eastern region which had vegetation and orange chromides while 13 did so where both were absent (Fig. 5). Obviously the number of breeding pairs in the two habitats did not differ. Nevertheless the pairs in the habitat without vegetation and orange chromides were reproductively more successful (Table 3, $X^2 = 4.30$; $p \leqq 0.05$).

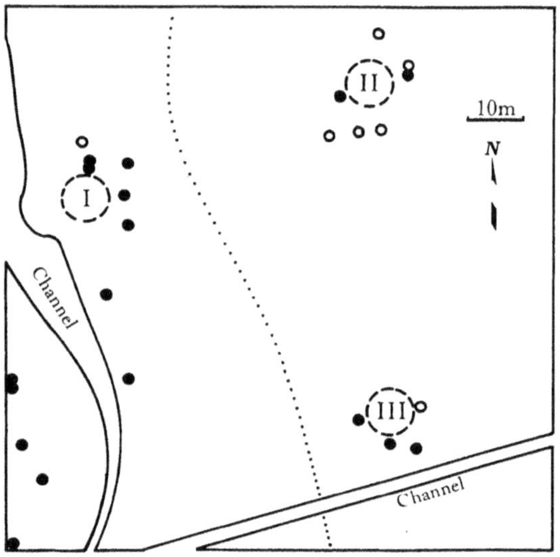

Fig. 5. E. suratensis nest dispersion during monsoonal dry season showing nests of successful (●) and unsuccessful (○) pairs. The dotted line separates the eastern region which had both rooted vegetation and *E. maculatus* from the western region which had neither. Roman numerals encircled by broken lines are brush-piles.

Nest location and time budgets

During the nesting phase members of an orange chromide pair alternately cared for offspring. The relieved partner foraged in the immediate vicinity or even wandered away from the nest; that parent often traveled several meters from the nest focus and sometimes browsed among non-breeding conspecifics. This behavior occupied 40–49% of the time budget (Fig. 6). A parent at the nest partitioned its time among tending, holding station, and aggressive interactions; the most important was holding station which occupied 40–47% of the time budget (Fig. 6). A parent holding station in sparse vegetation swam 10–20 cm above the nest and remained stationery; intruders were most often detected from here and then repulsed. Parents that constructed nests in dense vegetation were not as conspicuous in holding station as those in sparse vegetation.

Green chromides foraged fortuitously during parental care. Both parents guarded the nest at the same time as a team which accounted for 68–77% of the time budget (holding station, Fig. 6). Only 13–26% of time was occupied with wandering which rarely extended beyond the periphery of the nest – it never included foraging. Parents spent 6–10% of their time in interactions with intruders, a time allocation twice that for orange chromides (Fig. 6).

Table 3. Nests for *E. suratensis* during the monsoonal dry season.

Nest location	Number of nests	Number of spawn lost	Number of spawns reared to free swimming
Eastern region, vegetation + *E. maculatus*	11	5	6
Western region, no vegetation and no *E. maculatus*	13	1	12

Territoriality

Orange chromide pairs defended territories without distinct boundaries. Hypothetical boundaries calculated on the basis of mean distance between nest foci for colonies were drawn as circles in Figure 7 to contrast with the home ranges. A parent often swam as far as 2 m to display to an intruder. Only occasionally was a tending parent surprised by an intrusion in the nest focus. Conspecifics were actively repulsed while other fish, crabs and shrimps

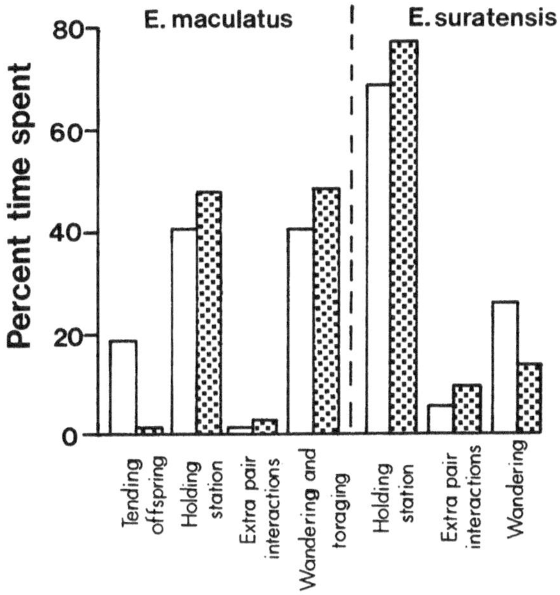

Fig. 6. Proportions of time spent in behavioral activities (time budgets) by *E. maculatus* and *E. suratensis*. Open bars represent the egg and stippled bars the eleutheroembryo phases of development of offspring.

were sometimes tolerated within 30 cm of the nest focus. At times a wandering parent traveled more than 3 m away from the nest focus and some remained away for as long as 5 min. Consequently within colonies there was considerable overlap in home ranges (Fig. 7).

The green chromide pairs defended nests with more precise boundaries; nest diameter was 50 cm to 2 m. The defended territories consisted of clearings in the substratum. All piscine intruders were repulsed. Even when unwary hermit crabs entered the territories they were picked up in the mouth of one of the parents and deposited outside the territory.

Egg-eleutheroembryo predation

Both orange and green chromide eggs and eleutheroembryos were subject to predation. The only obvious predators on orange chromide eggs and eleutheroembryos were other orange chromides. These cannibals were schools of mature adults (raiders) who were not breeding; raiding schools were composed of both males and females. The only predators observed on green chromide eggs and eleutheroembryos were also orange chromides. Here, the orange chromides were either paired and nesting nearby or raiding parties of non-breeding adults.

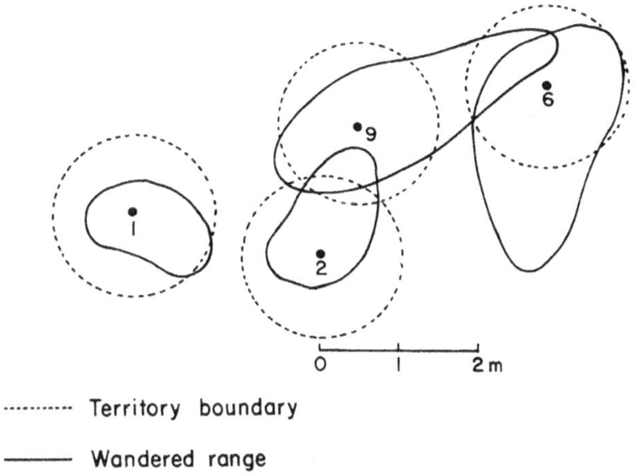

Fig. 7. Home ranges of four pairs of nesting *E. maculatus* in a colony. The numbers of the pairs are at the center of their hypothetical nesting territories shown as a broken line. Solid lines are areas in which members of a pair wandered during 15 min.

Discussion

The time of breeding for both species was the same, presumably because conditions in the two seasons favored survival of young. We conclude that the fast flowing murky waters of Negombo Lagoon during the monsoons are likely inimical to survival of chromide young, as has been reported in some other teleosts (Coulter 1970). Visual displays are important in the parental behavior of both chromide species (Cole & Ward 1969). The silt-laden murky waters of rainy seasons would restrict visual signalling and threaten survival of young.

Another factor favoring seasonal reproduction would be availability of food (zooplankton) for young. We did not collect data for zooplankton productivity in Negombo Lagoon, however, Sundararaj & Krishnamurthi (1975) reported that zooplankton was most abundant in Vellar estuary in South India when salinity was between 10–31 ppt, similar to that in Negombo Lagoon when chromides were breeding.

The other important determinant of reproductive success was the nesting habitat. Among more than a hundred nesting pairs of orange chromides observed only one constructed a nest at a site devoid of vegetation; it was within a brush pile. Pairs nesting after the nesting peak in the monsoonal season nested in dense vegetation and were successful presumedly because nests were well camouflaged. If offspring survival was the more important factor determining nest site selection we should have expected to observe all nests in dense vegetation. But since there was a distinct preference for sparse vegetation by orange chromides during breeding peaks we conclude that sparse vegetation was a compromise among factors favoring offspring survival as well as availability of food for foraging parents.

As the prominent consequence of choosing a nest where adult food was plentiful was the increased threat of cannibalism by non-breeding conspecifics, nesting success was largely dependent on the effectiveness of guarding. Orange chromides successfully repulsed raiding conspecifics as they searched a nest for the cryptic young, but early detection and preemptive attack were more effective. Parents and nests in sparse vegetation were relatively exposed but detection of predators was facilitated. Keenleyside & Bietz (1981) gave similar reason for nest site selection by the South American cichlid *Aequidens vittatus*.

Colonial nesting further enhanced nesting success of orange chromides in sparse vegetation. Davis (1978) stated that when nesting animals attack predators over a distance greater than the distance between nests then there will be more attackers within a colony than elsewhere and thereby colonial nesting will be selected for. Orange chromide parents attacked potential predators at distances greater than internest distances, but group attacks (mobbing) were never observed by us as reported for the Central American cichlids *Cichlasoma nigrofasciatum* and *Neetroplus nematopus* (McKaye 1977). However when intruders, usually conspecifics, were attacked by a member pair of a colony its neighbors sometimes held station above their own nests. Presumably neighbors thereby perceived potential danger. Perhaps colonial nesting is favored because it provides a mutually beneficial early warning system. Nesting in dense vegetation was a better tactic for orange chromides breeding after the peak presumably because few pairs nested then.

Ward & Wyman (1977) not only reported frequent cannibalistic raiders, but also some cannibalism by neighbors clustered into colonies. If this were a persistent habit it would select against coloniality unless under unusual conditions eggs and eleutheroembryos became necessary adult food. Ward & Wyman (1977) reported that the numbers of nesting pairs were so great that they could not be estimated. It seems likely that fish density increased cannibalism because the eggs and eleutheroembryos of neighbors became a necessary food resource. Fox (1975) in reviewing cannibalism states that increasing population density coupled with food scarcity can increase cannibalism.

The green chromide does not forage during parental care. Therefore its nest site selection was determined primarily by factors favoring offspring survival. This was effectively shown during the monsoonal season when reproductively more successful pairs nested in the region devoid of both orange chromides and vegetation. The green chromide feeds mainly on detritus and macrophytes.

Williams (1975) regarded sharing of guard duty among some cichlids as a serious evolutionary problem. Williams (1975) and Emlen (1973) concurred that a single parent could effectively guard a brood particularly when feeding offspring was not involved. There is ample evidence to support this view (Blumer 1979). *E. maculatus* and *E. suratensis* both display biparental care. In *E. maculatus* both parents contribute but they alternate so one parent actively defends the brood while the other is relieved to forage. Perrone & Zaret (1979) propose this is one situation that might favor biparental care since foraging keeps both members of the pair in good physical health for future guard duty. Many studies (Keenleyside & Bietz 1981, Keenleyside 1979, Perrone 1978) report that males and females behave differently in biparental species; generally males are more active. The considerable quantitative data undergoing analyses from this study does not yet suggest male and female orange chromides perform different roles, nor do sexual differences appear in laboratory studies (Allmand 1980, Cole & Ward 1969, Day 1975, Samarakoon 1975). Male and female of orange chromide pairs are equal in their ability to guard clutches but they do so only alternately – therefore the total parental investment is comparable to the full attention of just one parent. If the species must show parental care continuously for reproductive success, monogramy will be favored by the contraints of metabolic requirements. Finally, we will propose elsewhere that orange chromide pairs breed only once in a given season with optimal chance for reproductive success.

The reproductive strategy for *E. suratensis* is also biparental, but puzzling. Both parents are vigilant over their eggs and eleutheroembryos and neither forages (Fig. 4). Since there is no relief for either member of the pair their parental investment (at least in terms of time) is twice that of *E. maculatus*. The selective pressures which demand the constant and equal investment of both parents seem obscure at present. Green chromide parents spend twice as much time as orange chromide parents in interactions with intruders and repulsed more intruders. This extra effort hardly warrants the continued investment of both parents. We have considerably more data to evaluate from this field study and we hope to continue our field efforts in future. Perhaps then, we can better attempt an explanation of biparental care for *E. suratensis*.

Acknowledgements

The majority of this study was supported by NSF grant BNS76-01448 to JAW. We also thank Illinois State University for their research awards and preparation of the graphics.

References cited

Allmand, R. D. 1980. Comparative ethology of two color morphs of the Asian cichlid *Etroplus maculatus* (Pisces: Cichlidae). M.Sc. thesis, Illinois State University, Normal. 92 pp.

Balon, E. K. 1975. Terminology of intervals in fish development. J. Fish. Res. Board Can. 32: 1663–1670.

Barlow, G. W. 1981. Patterns of parental investment, dispersal and size among coral-reef fishes. Env. Biol. Fish. 6: 65–85.

Blumer, L. S. 1979. Male parental care in the bony fishes. Quart. Rev. Biol. 54: 149–161.

Borowsky, R. & J. Diffley. 1981. Synchronized maturation and breeding in natural populations of *Xiphophorus variatus* (Poeciliidae). Env. Biol. Fish. 6: 49–58.

Cole, J. E. & J. A. Ward. 1969. The communicative function of pelvic fin-flickering in *Etroplus maculatus* (Bloch). Behaviour 35: 179–199.

Coulter, G. W. 1970. Population changes within a group of fish species in L. Tanganyika following exploitation. J. Fish Biol. 2: 329–353.

Davis, N. B. 1978. Ecological questions about territorial behaviour. pp. 317–350. *In*: J. R. Krebs & N. B. Davies (ed.) Behavioural Ecology - an evolutionary approach, Blackwell, Oxford.

Day, R. L. 1975. Analysis of courtship behavior of the cichlid fish *Etroplus maculatus*. M.Sc. thesis, Illinois State University, Normal. 83 pp.

DeSilva, S. S. & E. I. L. Silva. 1979. Biology of young grey mullet, *Mugil cephalus* L., populations in a coastal lagoon in Sri Lanka. J. Fish Biol. 15: 9–20.

Dominey, W. J. 1981. Maintenance of female mimicry as a reproductive strategy in bluegill sunfish (*Lepomis macrochirus*). Env. Biol. Fish. 6: 59–64.

Emlen, J. M. 1973. Ecology: an evolutionary approach. Addison-Wesley, Reading. 493 pp.

Fox, L.R. 1975. Cannibalism in natural populations. Ann. Rev. Ecol. Syst. 6: 87–106.

Keenleyside, M. H. A. 1979. Diversity and adaptation in fish behaviour. Springer-Verlag, Berlin. 208 pp.

Keenleyside, M. H. A. & B. F. Bietz. 1981. The reproductive behaviour of *Aequidens vittatus* (Pisces, Cichlidae) in Surinam, South America. Env. Biol. Fish. 6: 87–94.

McKaye, K. R. 1977. Competition for breeding sites between the cichlid fishes of Lake Jiloa, Nicaragua. Ecology 58: 291–302.

Peiris, G. 1977. The physical environment. pp 3–30. *In*: K.M. de Silva (ed.) Sri Lanka - a survey, University Press of Hawaii, Honolulu.

Perrone, P., Jr. & T. M. Zaret. 1979. Parental care patterns of fishes. Amer. Nat. 113: 351–361.

Pritchard, D. W. 1967. What is an estuary: physical viewpoint. pp. 3–5. *In*: G. H. Luff (ed.) Estuaries, A. A. A. S., Washington, D.C.

Ridley, M. 1978. Paternal care. Anim. Behav. 26: 904–932.

Samarakoon, J. I. 1975. The role of behavioral experience and reproductive success in the cichlid fish *Etroplus maculatus* (Pisces: Cichlidae). M. Sc. thesis, Illinois State University, Normal. 113 pp.

Sundararaj, V. & K. Krishnamurthy. 1975. Nutrients and plankton: backwater and mangrove environments. pp 273–290. *In*: R. Natarajan (ed.) Recent Reseaches in Estuarine Biology, Hindustan Publ., Delhi.

Trivers, R. 1972. Parental investment and sexual selection. pp. 136–179. *In*: B. Campbell (ed.) Sexual Selection and the Descent of Man, Aldine, Chicago.

Ward, J. A. & R. L. Wyman. 1975. The cichlids of the resplendent isle. Oceans 8: 42–47.

Ward, J. A. & R. L. Wyman. 1977. Ethology and ecology of cichlid fishes of the genus *Etroplus* in Sri Lanka: preliminary findings. Env. Biol. Fish. 2: 137–145.

Williams, G. C. 1975. Sex and evolution. Princeton University Press, Princeton. 200 pp.

Env. Biol. Fish. Vol. 6, No. 1, pp. 105–109, 1981

Habitat segregation in a salt marsh among adult sticklebacks (Gasterosteidae)

John P. Worgan & Gérard J. FitzGerald

Département de biologie, Faculté des sciences et de génie, Université Laval, Québec G1K 7P4, Canada

Keywords:
Diel activity, Tidal creek, *Ruppia* pools, *Gasterosteus aculeatus*, *G. wheatlandi*, *Pungitius pungitius*, *Apeltes quadracus*, Interspecific competition

Synopsis

The spatial distribution, seasonal abundance and diel activity of four sticklebacks, *Gasterosteus aculeatus*, *G. wheatlandi*, *Pungitius pungitius*, and *Apeltes quadracus*, coexisting in a St. Lawrence salt marsh were examined to see how these closely related species share their habitat. While all four species breed in the Rivière des Vases, a tidal creek, only three species are found in the adjacent salt marsh pools, *A. quadracus* being absent. Results are interpreted in terms of avoidance of interspecific competition for space during the relatively short breeding season at this latitude.

Introduction

Habitat segregation among closely related species is thought to have evolved as a mechanism of avoiding or reducing interspecific competition for space. However, unless space is in short supply differences in its use are not indictive of ways of avoiding or reducing interspecific competition (Sale 1978).

The threespined, *Gasterosteus aculeatus*, the blackspotted, *G. wheatlandi*, the ninespined, *Pungitius pungitius* and the fourspined *Apeltes quadracus*, sticklebacks, migrate from the St. Lawrence estuary into rivers and tidal pools to breed from May to July. Unlike most estuarine fishes, sticklebacks are territorial during the breeding season and thus compete for space on breeding grounds. The purpose of the present study was to examine how these closely related species share space during their reproductive season.

Received 15.2.1980 *Accepted 27.8.1980*

Methods and material

The study was carried out at different stations along a 2.5 km section of Rivière des Vases, a small tidal creek which traverses a *Spartina* marsh before flowing into the St. Lawrence Estuary near Isle-Verte, Québec (Fig. 1) and in adjacent salt marsh pools. At high tide the channel ranged in width from 30 m at the mouth to 2.5 m in the upper portion and at low tide from 10 m at the mouth while remaining essentially unchanged upstream. Mid-channel depth ranged from 0.5 to 2.5 m depending upon tidal stage and distance from the creek mouth, where mean tidal amplitude was 2 m. Salinities over a tidal cycle usually varied from 3 ppt to 20 ppt at the mouth while 1.5 km upstream, the water was usually fresh. Water temperatures were

© Dr. W. Junk b.v. Publishers, The Hague. Printed in The Netherlands. 0378-1909/81/061-105 $01.00

similar among stations averaging 15.5° C (6–25°) during May and June and 19.5° C (15–26°) during July and August. The creek was frozen over from November to late April. Mean dissolved oxygen was 10.7 ppm (6.8–17.0), pH was usually 7.0 (6.5–8.0) and current speed was less than 0.3 $m \cdot s^{-1}$ with little variation among stations. Further details of water physiochemistry are given by Lavoie (1971). The creek's bed near the mouth was a sand-mud mixture, its banks largely mud; upstream the bed was sand. The only cover in the creek was several large boulders about 1 km from the mouth.

The salt marshes of the St. Lawrence Estuary, including our study site, are described by Reed & Moisan (1971). The tidal pools of the marsh are shallow (< 0.5 m), varying in area from 1 to 1000 m^2 and characterized by abundant growth of *Enteromorpha* sp., *Percursaria* sp. and *Ruppia maritima*. Most pools are in the *Spartina patens* littoral zone, are flooded regularly at spring tides and have a mean salinity of 24.1 ppt (19.1-26.7), while pools located further inland in the *Juncus* zone are rarely flooded and are essentially freshwater. Diel variation in dissolved oxygen, water temperature and salininty were measured every 2.4 h in four pools from June 18–28, 1979. Dissolved oxygen and temperature values attained mean minimum values (7.0 ppm, 14.0° C) just before dawn and mean maximum values (13.7 ppm, 22.3° C) in late afternoon. Salinity varied little (mean = 21.7 ppt; range = 19.2–24.3) as the pools were unflooded during the sampling period.

In 1978 and 1979 fish abundance in Rivière des Vases at each of four stations was estimated using two unbaited commercial minnow traps, one situated in the middle and the other near the bank at each station (Fig. 1). Traps were checked twice daily, at low and high tides, from May 5 to the end of August. Fish were counted and released. The reproductive status as described by Wootton (1976) was recorded for each; swollen abdomens for females and nuptial colours for males. To examine periods of peak abundance the total numbers of fish captured for each species for each two week period was expressed as a percentage of the total seasonal capture. Peak abundances of reproducing fish were similarly determined. Seining was carried out to confirm these peaks of abundance.

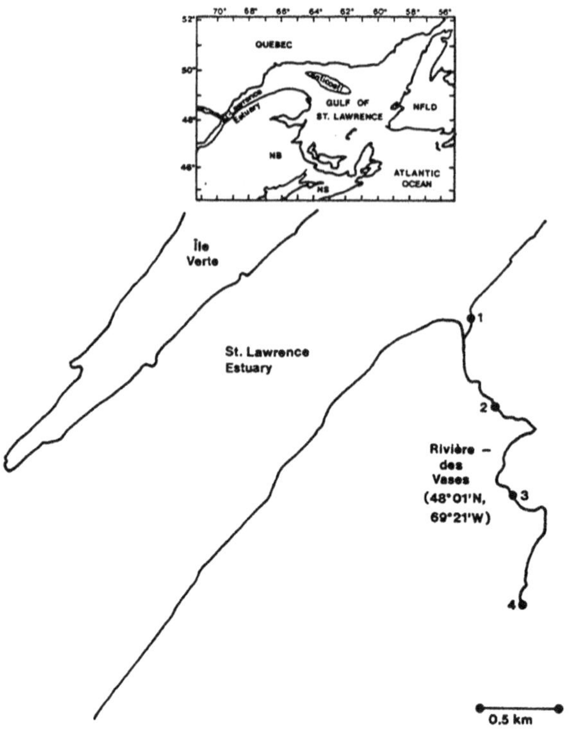

Fig. 1. Study region and position of stations in Rivière des Vases.

To evaluate seasonal changes in species abundance in the pools, three similar-sized pools (mean area = 115 m^2) bordering the estuary were thoroughly seined every two weeks during the 1979 season. As statistical analysis showed no differences in fish abundance among pools, the results were combined. We also seined on 17 June 1979 three similar sized pools at each of four locations at least 75 m apart at various distances from the estuary.

Diel activity was evaluated in the salt marsh over a nine day period from 18–28 June 1979 in four similar sized pools (mean area = 530 m^2). Fish were sampled (and released) every 2.4 h with two unbaited minnow traps per pool. As activity patterns (i.e. number of fish caught per trapping time) were similar statistically among pools, data were grouped and analysed by a Freidman non-parametric randomized block analysis of variance followed by the Nemenyi multiple comparisons test (Kirk 1968) to detect possible differences in diel activity among species.

Results and discussion

Abundance data for reproductive fish (Fig. 2) in Rivières des Vases showed that *G. aculeatus*, *G. wheatlandi* and *P. pungitius* finish spawning by 30 June but *A. quadracus* spawns until the end of July. By mid-July most *G. aculeatus* and *G. wheatlandi* have left the creek (Fig. 3) and there was a significant increase in the numbers of nonreproducing *A. quadracus* and *P. pungitius*. These two latter species occurred in the creek until freezeup in November. The only other fish species at the study site was the Atlantic tomcod, *Microgadus tomcod* (Lambert & FitzGerald 1979).

Species distribution along the creek showed *G. wheatlandi* were present only at station 1 with the exception of a few stray individuals captured at station 2 (Table 1). *Gasterosteus aculeatus* and *P. pungitius* were found at all four stations though were most abundant at station 1, while *A. quadracus* was present at station 1 to 3 but not at station 4.

Only three species occurred in the pools, *A. quadracus* being absent. Most reproduction took place in May and June for all species (Fig. 4). Pools were monitored after July 15 until freezeup but unlike in the creek there was no increase in the abundance of *P. pungitius*. All three species were most abundant in pools closest to the estuary (Table 2; Station A). Some *G. aculeatus* and *P. pungitius* occurred in freshwater pools (Station D) while *G. wheatlandi* did not occur at the freshwater sites.

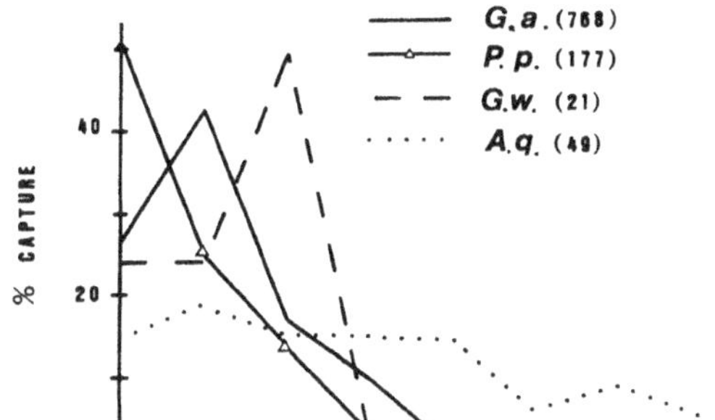

Fig. 2. Number of sexually mature fish captured in Rivière des Vases (1979) for each two week period expressed as a percentage of total seasonal capture. Figures in parentheses represent total number of fish caught throughout the season. *G.a.* = *Gasterosteus aculeatus*, *G.w.* = *G. wheatlandi*, *P.p.* = *Pungitius pungitius*, *A.q* = *Apeltes quadracus*.

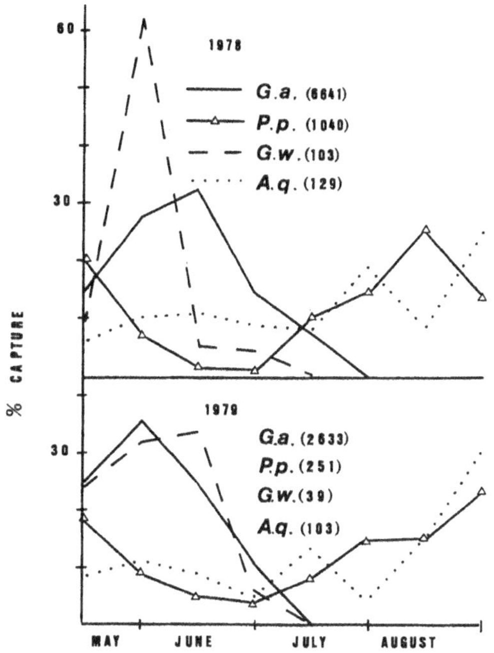

Fig. 3. Number of fish captured in Rivière des Vases for each two week period expressed as a percentage of total seasonal capture. Figures in parentheses represent total number of fish caught throughout the season. Symbols as in Figure 2.

The non-random distribution of the sticklebacks in the creek and pools can only partly be explained by species differences in water chemistry tolerances (Nelson 1968). For example, although we did not find *G. wheatlandi* in freshwater it can survive two weeks in a freshwater aquarium. *Apeltes quadracus*, although largely marine, occurs in inland lakes (Coad & Power 1973) and in salt marsh pools elsewhere (Rowland 1970). We introduced *A. quadracus* into a tidal pool; no mortality was

Table 1. Total number of fish captured at each station in Rivière des Vases during the spring and summer of 1978 and 1979.

	Station	1	2	3	4
	Year				
G. aculeatus	1978	3678	1108	889	766
	1979	1066	413	918	236
G. wheatlandi	1978	102	1	0	0
	1979	38	1	0	0
P. pungitius	1978	491	249	93	207
	1979	292	94	128	31
A. quadracus	1978	35	76	18	0
	1979	50	35	18	0

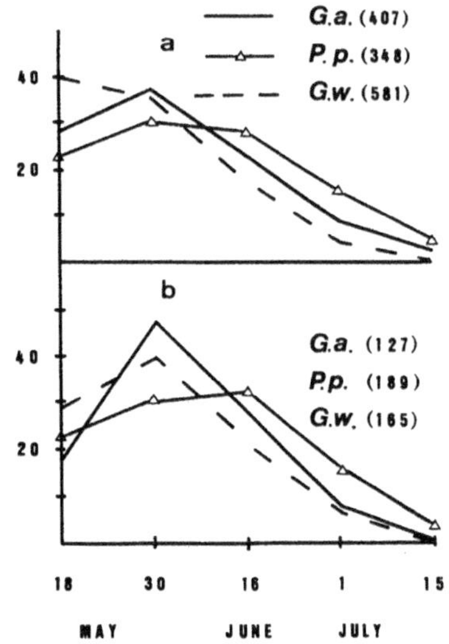

Fig. 4. Number of fish captured bimonthly in pools expressed as a percentage of total seasonal capture. (a) represents total numbers of fish and (b) numbers of sexually mature fish. Symbols as in Figure 2.

Table 2. Number of fish in pools on June 17, 1979. Values are total of three pools at each station.

	Station			
	A*	B	C	D**
G. aculeatus	48	10	47	2
G. wheatlandi	114	39	15	0
P. pungitius	64	8	14	8

* Stations A-D were 25, 100, 175 and 275 m from the estuary respectively.
** Freshwater station.

observed after one week but we do not know if they can reproduce there. The absence of *A. quadracus* from our pools is curious because this species coexists with *G. wheatlandi* and *G. aculeatus* in similar pools near Long Island, New York (Rowland 1970).

The above suggests that it is excluded from the pools by *P. pungitius*. This hypothesis is supported by laboratory observations where *P. pungitius* harrassed and bit off the tails of *A. quadracus* when placed together in aquaria. Furthermore Wootton (1976) reports that fecundity is lower and parental care less well developed for *A. quadracus* than for the other species. Perhaps *A. quadracus* competes poorly with or inadequately protects its eggs from the other species during the breeding season. The extended reproductive period of *A. quadracus* possibly allows repeat spawning thus reducing competition for space with other species.

Due to the climatic conditions at this northern latitude it is not surprising that there is considerably more overlap in peaks of abundance of sticklebacks at Isle-Verte than among those at Long Island (Rowland 1970).

Pool fish were essentially diurnal with only 1.3% of 2985 *G. aculeatus*, 5.7% of 247 *G. wheatlandi* and 6.8% of 813 *P. pungitius* captured at night. Activity for all species was uniformly distributed between dawn and dusk and there were no significant differences in activity among species.

Although little is known about stickleback diel activity in intertidal areas Manzer (1976) reported that lacustrine *G. aculeatus* displayed a bimodal dawn-dusk feeding activity pattern while Girsa (1975) reported that immature *G. aculeatus* in the White Sea were arthythmic under Arctic summer conditions.

Our data are suggestive of interspecific competition for space. Experiments to detect possible interspecific competition for food are in process.

Acknowledgements

This study was supported by a N.S.E.R.C. Grant AR-6179-1375 to G.J.F. We thank Gordon Walsh for field assistance and Jeremy McNeil for critically reading the manuscript.

References cited

Coad, B. W. & G. Power. 1973. Life history notes and meristic variation in the fresh water four spine stickleback, *Apeltes quadracus* (Mitchell), near Sept-Isles, Quebec. Naturaliste can. 100: 247–251.

Girsa, I. I. 1975. The diurnal feeding rhythm in the feeding of underyearlings of some White Sea fish at different photoperiods. J. Ichthyol. 102–109.

Kirk, R. E. 1968. Experimental design: procedures for the behavioral sciences. Brooks & Cole Publishers, Belmont. 577 pp.

Lambert, Y. & G. FitzGerald. 1979. Summer food and movements of the Atlantic tomcod *Microgadus tomcod* (Walbaum) in a small tidal creek. Naturaliste can. 106: 555–559.

Lavoie, R & G. Beaulieu. 1971. Salinité des eaux de surface du St-Laurent. Naturaliste can. 98: 191–193.

Manzer, J. I. 1976. Distribution, food and feeding of the threespine stickleback, *Gasterosteus aculeatus*, in Great

Central Lake, Vancouver Island, with comments on competition for food with juvenile sockeye salmon, *Oncorhynchus nerka*. U.S. Fish Bull. 74: 647–668.

Nelson, J. S. 1968. Salinity tolerance of brook sticklebacks, *Culaea inconstans*, freshwater ninespine sticklebacks, *Pungitius pungitius* and freshwater four spine sticklebacks, *Apeltes quadracus*. Can. J. Zool. 46: 663–667.

Reed, A. & G. Moisan. 1971. The *Spartina* tidal marshes of the St. Lawrence estuary and their importance to aquatic birds. Naturaliste can. 98: 905–922.

Rowland, W. J. 1970. Behavior of three sympatric species of sticklebacks and its role in their reproductive isolation. Ph.D thesis, State University of New York at Stony Brook. 170 pp.

Sale, P. F. 1978. Habitat partitioning and competition in fish communities. pp. 323–331. *In*: R. H. Stroud & H. Clepper (ed.) Predator-Prey Systems in Fisheries Management, Sport Fishing Institute, Washington D.C.

Wootton, R. J. 1976. Biology of the sticklebacks. Academic Press, London. 387 pp.

Env. Biol. Fish. Vol. 6, No. 1, pp. 111–121, 1981

Factors affecting fish diversity on a temperate reef

John S. Stephens, Jr. & Kim E. Zerba

Department of Biology, Occidental College, Los Angeles, California 90041, U.S.A.

Keywords:
Ecotone, Non-equilibrium, Niche, Reproductive success, Predators, Resource utilization

Synopsis

Subtidal studies of fish stocks conducted along a rock breakwater in Southern California reveal a rich and diverse fauna with elements from cool and warm temperate regions. More than 100 species have been observed with half of these resident. Only a single large, permanently territorial species occurs on the reef though many small, often cryptic species defend home and/or reproductive sites. Many of the larger, mobile species orient to temperature stratification and a portion of the diversity can be allied to the ecotonal effect of stable stratification. Both large predators and herbivores are rare; the majority of the species are micropredators, grazing on sedentary or benthic microinvertebrates.

Reproduction is keyed to spring-summer seasons, generally with pelagic eggs, embryos and larvae. The live-bearing resident embiotocids show the least fluctuation in numbers while many oviparous species have shown major changes in annual success during the study period. Although the success of individual species varies, the diversity and total abundance of fishes has remained relatively stable for the reef community as a whole. This stability cannot be considered to reflect an equilibrium community and in fact the structure of the community has changed significantly during the five year study.

Introduction

In recent years, the question of community diversity has been the subject of field research as well as a source for theoretical models of ecology and evolution. Fish communities, particularly tropical reef communities, have been used as examples of high diversity, but these have yielded conflicting interpretations (Smith & Tyler 1972, Roughgarden 1974, Sale 1977). Most of these discussions have used competitive equilibrium models though there is little evidence that equilibria occur (Houston 1979). At present, little information has been presented with regard to temperate reef fish diversity though Sale (1977) compared tropical and temperate reefs using available published data. With the exception of some short term experimental recruitment studies (Sale & Dybdahl 1975, Molles 1978, Talbot et al. 1978) the analyses have not included community changes with time. Further, the methodologies employed in diversity sampling and analysis have not allowed unbiased repetitive sampling. In this study we report fish stock data collected over a five-year period (1974–79) using non-destructive sampling techniques.

Received 26.2.1980 Accepted 29.8.1980

© Dr. W. Junk b.v. Publishers, The Hague. Printed in The Netherlands. 0378-1909/81/061-111 $02.20

Materials and methods

The non-cryptic fish species in King Harbor, California have been counted by diver transect methods since 1974. The methodology utilized (Terry & Stephens 1976) involves five minute isobathic transects swum by divers at 5 m depth intervals along a rocky breakwater. Two divers run each transect and their results are compared, duplicate transects are run at each depth. King Harbor, located at the southern end of Santa Monica Bay, is influenced by hot water effluent from Southern California Edison's steam elecrical generating plant and by upwelling from the adjacent Redondo submarine canyon. The effects of these environmental factors on fish diversity have been discussed previously (Terry & Stephens 1976, Stephens & Ellison 1977, Stephens & Palmer 1979). For the first 18 months of the study, data were recorded monthly from 17 sets of duplicate transects at the principal study site (the outer terminus of the breakwater) with a total of 27 monthly duplicate sets within the harbor habitat. Comparative transects were run at several similar sites outside of the harbor. Since April 1976, the transects have been quarterly. A total of 357 transects were completed from 1974–78 at the principal study site.

The cryptic fish fauna has been examined by intensive diver surveys and recently by the use of quinaldine to collect and return m^2 såmples of fishes from benthic epifaunal growth. Fish eggs, embryos and larvae have been monitored in the water column using quantitative meter net tows (Ehrlich et al., unpubl. data). Additionally, food studies have been completed on many important harbor species (Stephens & Ellison 1977, Ellison et al. 1979). Available food resources have been analyzed by Straughan (unpubl. data). Some preliminary observations of fish activity and behavior have been carried out in field and aquarium.

All data are key-punched, stored on tape, and analyzed by computer. We are using number of species as the measure of diversity (Poole 1974). Abundance comparisons were made for all common species using the general linear model analysis of variance procedure including all possible linear contrasts. Rank comparisons were made using a rank index, the index of community importance. This index is derived by summing the rank positions of mean abundance/transect and fidelity (% occurrence) for each species and reranking the sums. This statistic resulted from comparing these results for a given year with a number of alternative data rankings, and it appears to give a ranking that agrees closely with our observed estimates of relative importance. Significance was tested with Kendall's tau rank association statistic.

Results and discussion

Table 1 presents a list of 105 fish species observed during 667 isobathic transects conducted at the King Harbor Breakwater since 1974. Each five-minute transect averaged 11.8 species and 238 individuals which is significantly higher than the comparative transects at Palos Verdes, Santa Monica Harbor and Catalina Island. The mean diversity per station (43.8 species) at King Harbor is more than double that at Palos Verdes (19.6) or Santa Monica Harbor (17.0) (Stephens & Palmer 1979). In addition to these observational data, in 39 quinaldine samples at King Harbor, a mean of 2.4 species and 13.7 cryptic individuals per m^2 was found within the epibenthic substrate. The high diversity at King Harbor is largely due to thermal stratification (Terry & Stephens 1976) and the ecotonal nature of the study area. The term 'resident' (Table 1) refers to the fact that the species regularly occurs within the study area. It does not indicate that population turn-over is not occurring. Tagging studies have confirmed the regular occurrence of certain individuals within the habitat, but the regular occurrence of wandering schools (i.e. *Atherinops*, *Hyperprosopon*, etc.) suggests continual exchange of individuals with surrounding habitats. Seasonal fishes occur regularly during specific months while transient species occur sporadically but often in large numbers and unusual fishes are rarely seen in the study area.

Table 2 presents a five-year summary of data for all breakwater rock substrate stations within King Harbor. These data represent means of means and are approximately normal in distribution. The principal study area is located at the tip of the breakwater at the harbor entrance and is represent-

Table 1. Species of fish observed in King Harbor by Occidental divers (1974–79).

Species	Status *	Species	Status *
Agonopsis sterletus	U	*Neoclinus stephansae*	R
Amphisticus argenteus	U	*Neoclinus uninotatus*	R
Anchoa delicatissima	U	*Odontopyxis trispinosa*	U
Anisotremus davidsonii	T	*Ophiodon elongatus*	S
Artedius creaseri	R	*Orthonopias triacis*	R
Artedius notospilotus	R	*Oxyjulis californica*	R
Atherinops affinis	R?	*Oxylebius pictus*	R
Atherinops californiensis	R?	*Paraclinus integripinnis*	R
Atractoscion nobilis	U	*Paralabrax clathratus*	R
Brachyistius frenatus	R	*Paralabrax maculatofasciatus*	R
Caulolatilus princeps	S	*Paralabrax nebulifer*	R
Cephaloscyllium ventriosum	S	*Paralichthys californicus*	S
Cheilotrema saturnum	R	*Parophrys vetulus*	S
Chromis punctipinnis	R	*Phanerodon furcatus*	R
Citharichthys stigmaeus	R	*Platyrhinoidis triseriata*	S
Clevelandia ios	R	*Pleuronichthys coenosus*	R
Clinocottus analis	R	*Pleuronichthys decurrens*	R
Coryphopterus nicholsii	R	*Pleuronichthys ritteri*	R
Cymatogaster aggregata	S	*Pleuronichthys verticalis*	U
Embiotoca jacksoni	R	*Porichthys myriaster*	U
Embiotoca lateralis	U	*Quietula y-cauda*	R
Engraulis mordax	S	*Rathbunella hypoplecta*	R
Genyonemus lineatus	S	*Rhacochilus toxotes*	R
Gibbonsia elegans	R	*Rhacochilus vacca*	R
Gibbonsia metzi	R	*Rhinobatos productus*	U
Girella nigricans	R	*Sarda chiliensis*	S
Gobiesox rhessodon	R	*Scomber japonicus*	T
Gymnothorax mordax	U	*Scorpaena guttata*	R
Halichoeres semicinctus	R	*Scorpaenichthys marmoratus*	R
Hermosilla azurea	T	*Sebastes auriculatus*	R
Heterodontus francisci	U	*Sebastes carnatus*	R
Heterostichus rostratus	R	*Sebastes dalli*	R
Hexagrammos decagammus	S	*Sebastes miniatus*	S
Hippoglossina stomata	S	*Sebastes mystinus*	R
Hyperprosopon argenteum	R?	*Sebastes paucispinis*	S
Hypsoblennius gentilis	R	*Sebastes rosaceus*	U
Hypsoblennius gilberti	R	*Sebastes serranoides*	R
Hypsoblennius jenkinsi	R	*Sebastes serriceps*	R
Hypsopsetta guttulata	R	*Sebastes vexillaris*	R
Hypsurus caryi	R	*Semicossyphus pulchrum*	R
Hypsypops rubicundus	R	*Seriola dorsalis*	T
Ilypnus gilberti	R	*Seriphus politus*	T
Leiocotius hirundo	U	*Sphyraena argentea*	T
Leuresthes tenuis	R?	*Squalus acanthias*	T
Lythrypnus dalli	R	*Squatina californica*	T
Lythrypnus zebra	R	*Syngnathus* sp.	R
Medialuna californiensis	R?	*Synodus lucioceps*	U
Menticirrhus undulatus	T	*Torpedo californica*	U
Micrometrus minimus	R	*Trachurus symmetricus*	T
Mola mola	U	*Urolophus halleri*	R
Mugil cephalus	U	*Xenistius californiensis*	U
Myliobatis californica	R	*Xystreurys liolepis*	R
Neoclinus blanchardi	R		

* R = resident, S = seasonal, T = transient, U = rare

Table 2. Annual data for diversity and density of fishes in the King Harbor.

	A. Number of species per station per year			B. Number of specimens per station per year		
	Stations			Stations		
Years	3	1	2	3	1	2
1974–75	11.7	13.6	14.2	345	274	333
1975–76	11.9	13.1	11.2	340	289	112
1976–77	11.6	10.4	9.3	449	302	146
1977–78	12.0	10.5	10.6	190	226	263
1978–79	12.0	11.5	10.8	525	327	241

ed by Stations 1 and 3, while Station 2 is located more deeply within the harbor. If these data are compared by a two-way analysis of variance, no significant difference is observed for either the species per station per year or the fish per station per year. This consistency of data suggests that the fish populations on the King Harbor reef are stable (in equilibrium). These data do not, in fact, reflect such stability. They may reflect 'niche stability' rather than species stability.

Figures 1–6 display the annual fluctuations in mean daily abundance per transect of important species. They are grouped by changes in their annual abundance characteristics: Figures 1 and 2 represent species that are relatively stable; Figure 3 – increasing; Figure 4 – decreasing; Figures 5 and 6 – non-directionally fluctuating species. Table 3 presents data from linear contrasts between years, indicating where significant differences in abundances were observed.

The data for 1978 on these graphs is biased due to the absence of winter transects, a period when fish abundance is usually high. Stormy weather during the sampling period precluded collection of these data. Our 1979 transects appear to show a reversal of the downward trend observed in the incomplete data of 1978 for the embiotocids (Fig. 1, 2). The data in Figure 4 represent successful breeding and recruitment to the harbor for *Hypsypops* and *Semicossyphus*, especially during 1976, while the decrease in abundance of *Citharichthys* (Fig. 4) was allied (Ehrlich et al. 1979) to an increase in bottom water temperatures. These directional changes in species abundance were noted throughout the

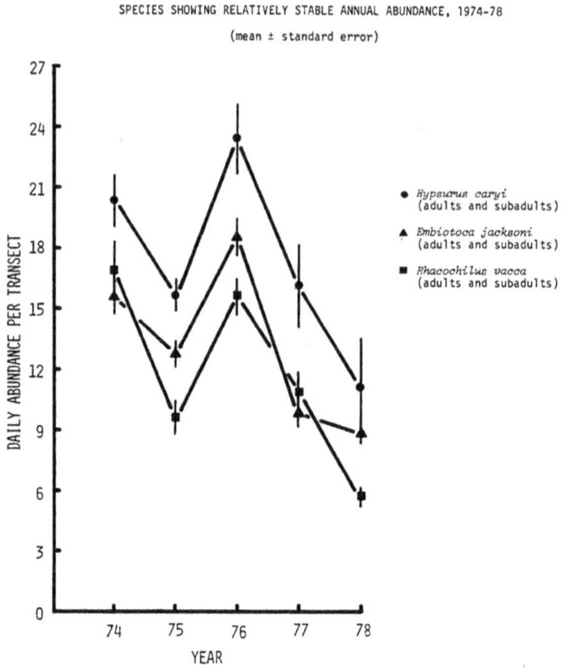

Fig. 1. Annual abundance of *Embiotoca jacksoni*, *Hypsurus caryi* and *Rhacochilus vacca* in the study area.

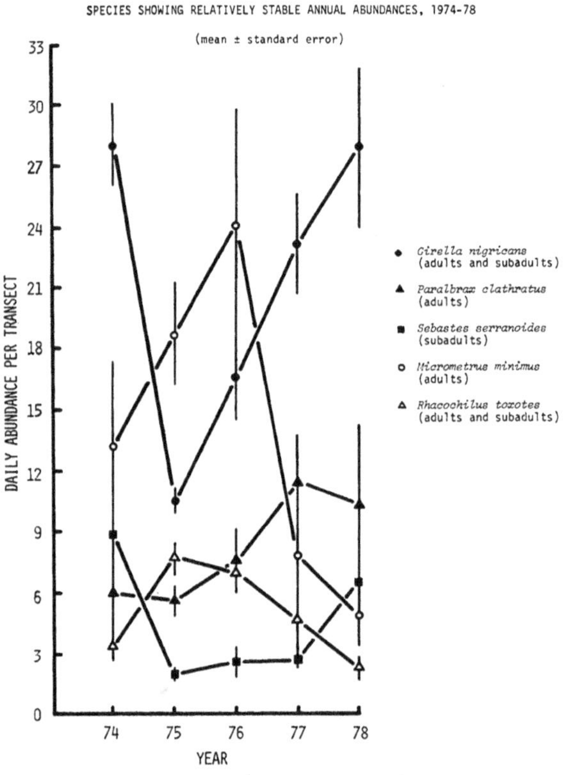

Fig. 2. Annual abundance of *Girella nigricans*, *Micrometrus minimus*, *Paralabrax clathratus*, *Rhacochilus toxotes* and *Sebastes serranoides* in the study area.

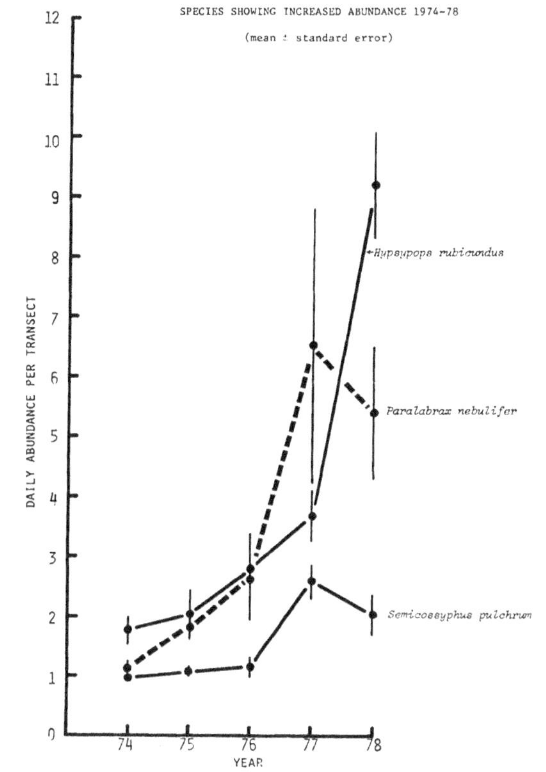

Fig. 3. Increasing abundance of *Hypsypops rubicundus*, *Paralabrax nebulifer* and *Semicossyphus pulchrum* in the study area.

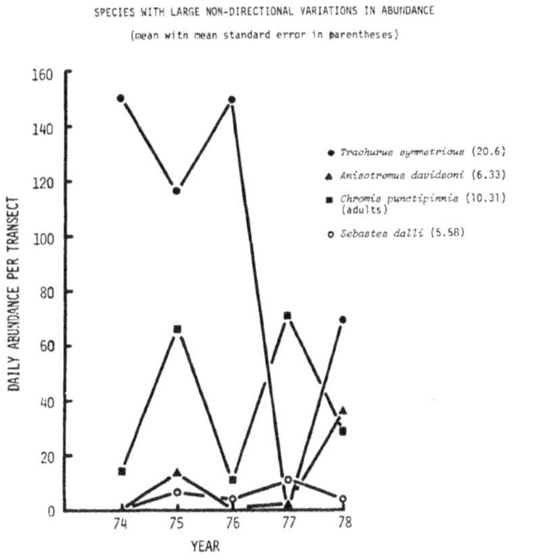

Fig. 5. Fluctuations in abundance of *Anisotremus davidsonii*, *Chromis punctipinnis*, *Sebastes dalli* and *Trachurus symmetricus* in the study area.

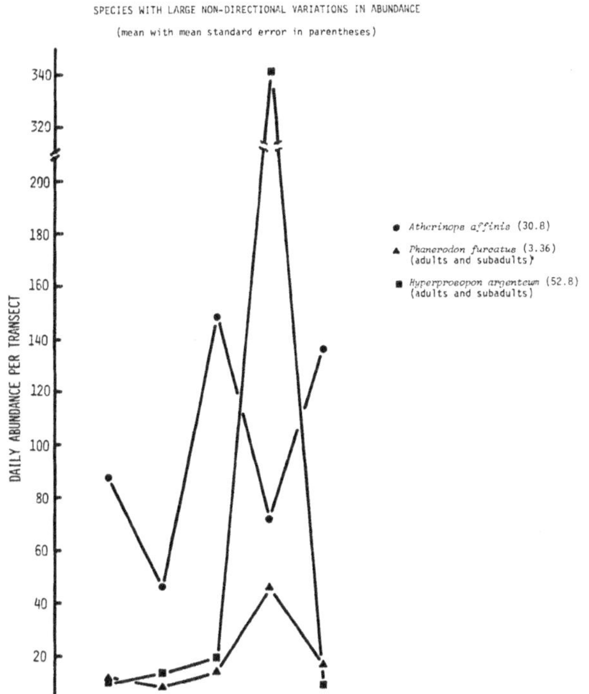

Fig. 6. Fluctuations in abundance of *Atherinops affinis*, *Hyperprosopon argenteum* and *Phanerodon furcatus* in the study area.

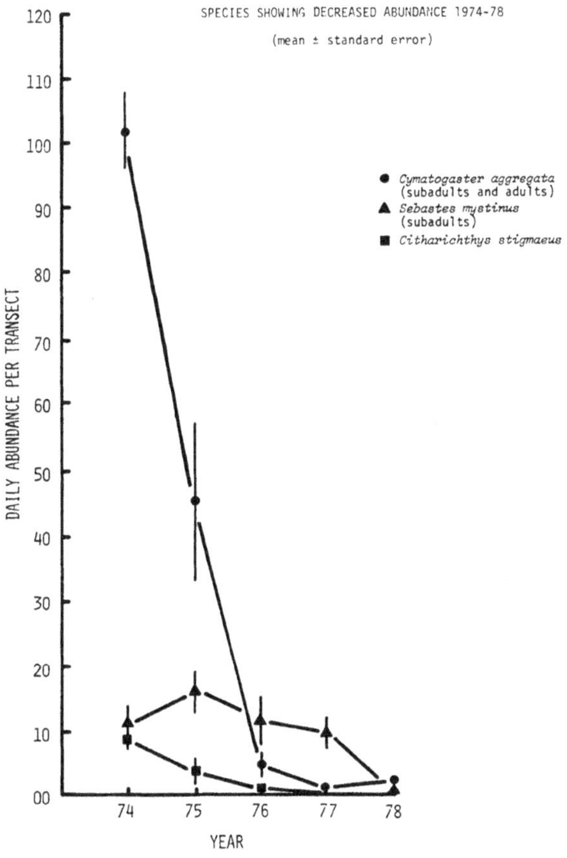

Fig. 4. Decreasing abundance of *Citharichthys stigmaeus*, *Cymatogaster aggregata* and *Sebastes mystinus* in the study area.

Table 3. Years when species abundance differs significantly as measured by a linear contrast (p = 0.05).

Species	Figure #	Comparisons
Atherinops affinis	5	1974–5,6; 1975–6,7; 1976–7
Chromis punctipinnis	5	1975–6; 1976–7
Citharichthys stigmaeus	4	1974–5
Cymatogaster aggregata	4	1974–5
Hyperprosopon argenteum	5	1977–all years
Hypsypops rubicundus	3	1974–8,9; 1975–7,8,9; 1976–8,9: 1977–8
Paralabrax nebulifer	3	1974–7,8; 1975–7,8; 1976–7
Semicossyphus pulchrum	3	1975–7,9; 1976–9
Trachurus symmetricus	5	1978–4,5,7

Southern California Bight. The species of embiotocids undergoing major population fluctuations (*Cymatogaster*, Fig. 4; *Hyperprosopon* and *Phanerodon*, Fig. 6) are all schooling, wandering or seasonally migrating species. Similarly, *Trachurus*, *Anisotremus* (Fig. 5) and *Atherinops* (Fig. 6) represent schooling species not tied to the substrate. By contrast, the variations in *Chromis* and *Sebastes dalli* probably represent variations in reproductive success.

It is obvious from these data there has been considerable fluctuation in abundance of individual species on the reef even though the numbers of species and number of individuals have remained stable.

Changes in the composition of a fish community can be evaluated if a rank order relationship can be established for community members. These ranks at successive sampling periods can then be compared as a measure of temporal stability on the reef using Kendall's tau measure of rank association. The difficulty with this process is in selecting criteria by which to rank the species. Neither abundance nor fidelity rankings appear adequate alone, and we have used a combined index of community importance as our ranking measure. Table 4 presents comparative ranks for five completed study years (1974–78). The comparison between 1974 and 1978 data does not show significant association at 0.05, but the degree of association decreases regularly with time (r = 0.94). A major turn-over in fish fauna is suggested by the lack of significant association between years 1 and 5 of the study. These indicators demonstrate statistically that the fish stocks are not in equilibrium but are changing.

Having presented data strongly suggesting that the fish stock at King Harbor is not stable but that the diversity of the stock remains constant and complex, the mechanism for maintenance of diversity should be explored. The primary cause of diversity within the King Harbor habitat is described as ecotonal due to the continual presence of cool temperate and warm temperate water. The cool waters upwelling from adjacent Redondo submarine canyon are entrained by power plant systems, and the warm, stable surface waters are affected by both insolation and hot water discharge. The depth relationships of fishes largely reflect temperature preferences which have been documented by laboratory gradient tank studies (Ehrlich et al. 1979, Shrode unpublished). During most of the year the thermocline occurs above 9 m and the fishes below that depth either prefer cooler water or soft substrate-rock sand interface habitat or both. Generally, the rockfishes of the genus *Sebastes*, *Oxylebius pictus*, *Coryphopterus nicholsii*, *Paralabrax nebulifer*, and the surfperches (Embiotocidae) *Hypsurus caryi*, *Phanerodon furcatus*, *Rhacochilus vacca* and *R. toxotes* all prefer cooler waters. Other species, especially those in shallower positions, are thermally tolerant with some actively preferring warm water (i.e. *Atherinops*, *Girella*). To some degree the diversity at King Harbor represents the presence of two separate thermally distinct communities.

In a fish stock with 'enhanced' biomass and diversity, a classical interpretation might suggest increased species packing through niche specialization. Our observations suggest that little specialization occurs. Niche specializations primarily develop in two areas: habitat and food resource utilization. Habitat specializations are often indicated behaviorally by territorial defense. With the exception of small cryptic benthic species (blenniids, clinids, cottids, gobies, etc.), the only permanently territorial species at King Harbor is the garibaldi, *Hypsypops rubicundus* (Clarke 1970).

Most of the epibenthic species do not seem to maintain any diel home position along the reef,

Table 4. Species ranks by index of community inportance, 1974–78.

1974	1975	1976	1977	1978
1. *Atherinops affinis*	5.0	1.5	15.5	15.0
2. *Rhacochilus vacca a*	9.5	1.5	7.0	12.0
3. *Hypsurus caryi a*	2.0	4.0	6.0	8.0
4. *Embiotoca jacksoni a*	4.0	9.0	10.0	7.0
5. *Phanerodon furcatus* a	15.5	19.0	18.0	9.0
6. *Medialuna californiensis*	23.5	15.0	15.5	17.0
7. *Paralabrax clathratus* a	11.0	11.0	4.0	5.0
8. *Oxyjulis californica* s	1.0	3.0	2.0	1.0
9. *Cymatogaster aggregata* a	9.5	37.0	39.0	34.0
10. *Citharichthys stigmaeus*	34.0	—	—	—
11. *Girella nigricans* a	22.0	20.0	13.0	13.0
12. *Gibbonsia elegans*	28.0	39.0	28.0	20.0
13. *Sebastes mystinus* s	21.0	18.0	23.0	—
14. *Trachurus symmetricus*	14.0	30.5	—	28.0
15. *Micrometrus minimus*	3.0	6.0	12.0	16.0
16. *Coryphopterus nicholsii*	24.0	33.0	22.0	29.0
17. *Heterostichus rostratus* a	33.0	26.5	29.0	27.0
18½ *Rhacochilus toxotes* a	19.0	14.0	27.0	24.5
18½. *Pleuronichthys coenosus*	25.0	24.0	33.5	31.0
20 *Sebastes serranoides* s	27.0	21.0	26.0	14.0
21. *Cymatogaster aggregata* s	17.5	—	—	—
22. *Scorpaenichthys marmoratus*	35.0	25.0	37.0	—
23. *Caulolatilus princeps*	8.0	28.0	30.0	39.5
24. *Embiotoca jacksoni* s	6.0	5.0	5.0	6.0
25. *Clinocottus analis*	32.0	—	31.5	35.5
26. *Paralabrax maculatofasciatus* a	37.0	38.0	33.5	18.0
27. *Oxylebius pictus*	15.5	13.0	19.0	26.0
28. *Hyperprosopon argenteum* a	20.0	30.5	20.0	30.0
29. *Chromis punctipinnis* a	13.0	8.0	1.0	2.0
30. *Hypsurus caryi* s	12.0	7.0	11.0	33.0
31. *Rhacochilus vacca* s	7.0	10.0	9.0	11.0
32. *Scorpaena guttata*	—	32.0	39.0	39.5
33. *Orthonopias triacis*	—	—	—	—
34. *Paralabrax nebulifer* a	29.0	22.0	14.0	10.0
35. *Halichoeres semicinctus*	26.0	23.0	17.0	24.5
36. *Sebastes paucispinis*	—	—	—	—
37. *Lythrypnus dalli*	—	—	39.0	—
38. *Girella nigricans* s	17.5	16.0	3.0	3.0
39. *Hyperprosopon argenteum* s	36.0	35.0	—	—
40. *Hypsypops rubicundus* a	31.0	12.0	8.0	4.0
41. *Phanerodon furcatus* s	23.5	17.0	25.0	19.0
42. *Semicossyphus pulchrum* a	38.0	36.0	21.0	23.0
44½ *Sebastes dallii*	—	29.0	24.0	21.5
44½ *Anisotremus davidsonii*	—	—	31.5	21.5
44½ *Rhacochilus toxotes* s	—	26.5	39.0	—
44½ *Cheilotrema saturnum* a	—	—	—	32.0

Kendall's tau analysis of association by year ($\alpha = 0.05$).

Contrasts	P	Results
1974 with 1975	0.001	significantly associated
1974 with 1976	0.01	significantly associated
1974 with 1977	0.05	significantly associated
1974 with 1978	>0.05	not significantly associated
1975, 1976 & 1977 all <0.01		

moving and feeding within their general depth level. Few of the fishes also appear to have specialized nocturnal positions, usually resting on or above the substrate in about the same level as during the day. *Chromis punctipinnis*, a diel planktivore, does move on and off the reef in a nocturnal-diurnal cycle and the wrasses (Labridae) show some nocturnal specialization, but all of the above exceptions are tropical offshoots. Ebeling & Bray (1977) have discussed this difference between tropical and temperate reefs. The lack of nocturnal specialization is exemplified by female *Cymatogaster aggregata* (Embiotocidae) which school and feed in midwater during the day and descend to the open sandy substrate at night where they are often observed laying on their sides and are quiescent enough to be handled by divers. Such an 'adaptation' suggests low levels of nocturnal predation.

Bray & Ebeling (1975), Stephens & Ellison (1977), Ellison et al. (1978), and Love & Ebeling (1978) have recently discussed food utilization by southern California rocky shore fishes. In general, there are few herbivores (*Medialuna*, *Girella*, and *Hermosilla*), a few resident or seasonal planktivores (*Chromis*, *Cymatogaster*, *Atherinops*, etc.), and a limited group of resistent macropredators (serranids, scorpaenids). The majority of the species are epifaunal grazers or pickers (micropredators). There are a few food specialists among this group (i.e. *Rhacochilus vacca*, which is malacophagous), but most of the dietary differences between these species probably reflect particle size selection and depth differences of the foraging zones.

The lack of habitat or food specialization in reef fishes was also suggested by Sale (1977) but he invoked living-space competition and high predation levels as a lottery which played the major role in diversity maintenance. In King Harbor space does not seem to be an important competitive consideration for epibenthic species. During the summer of 1978 divers collected data on agonistic encounters observed between fishes and attempted to develop data on home range or territory boundaries. Table 5 presents the results of four days (approximately 16 hours) of observations. Agonistic encounters are almost nonexistent. These data agree with our general impression of little interspecific antagonism among these reef species.

		Hypsypops rubicunda A	*Hypsypops rubicunda* S	*Embiotoca jacksoni* A	*Embiotoca jacksoni* S	*Embiotoca jacksoni* J	*Rhacochilus vacca* A	*Rhacochilus vacca* S	*Rhacochilus vacca* J	*Hypsurus caryi*	*Paralabrax clathratus*	*Micrometrus minimus*	*Semicossyphus pulchrum*	*Chromis punctipinnis*	*Girella nigricans*	*Oxyjulis californica*	*Halichoeres semicinctus*	*Scorpaenichthys marmotatus*	Total
Hypsypops rubicunda	A	13		2									1		1		1		18
	S		1					4								1			6
Embiotoca jacksoni	A			4						1									5
Rhacochilus vacca	A						1												1
	S							1											1
	J																		
Hypsurus caryi	A				1					2									3
Paralabrax clathratus	A										1								1
Micrometrus minimus	A											11							11
Semicossyphus pulchrum	S													1					1
Girella nigricans	S													1					1
Gibbonsia metzi	A				1		1											1	3
Total																			51

A = adult, S = subadult, J = juvenile

Table 5. Agonistic encounters observed through field observations (1977–78).

Spatial competition is refuted by the lack of territoriality as well as the lack of aggressive behavior. All rationale for territorial behavior requires competition for some limiting resource (Verner 1977).

The degree of heterotypic schooling (Table 6) is also indicative of a low level interspecific aggression. Many of these heterotypic schools involve fishes utilizing the same food resources. Heterotypic schools can be advantageous under extreme levels of predation (increased school size; Ehrlich & Ehrlich 1973) or in the presence of abundant but patchy food resources.

Our observations suggest that predation upon adult fishes in King Harbor is at a very low level. There are few common predators large enough to ingest even moderate sized actinopterygians. This lack of large predators may be the result of long term selective fishing pressure in a metropolitan environment. The presence in the area of surviving injured or diseased fish also attests to the reduced level of predatory selection

By contrast, the availability of food resources, especially microcrustacea, suggests that food supplies are not limiting. In a series of caging studies carried out during 1977 an average standing crop of 8000 microcrustacea per m^2 was observed in the samples. When samples were scraped they regained their former biomass within three months even with grazing pressure (uncaged). Similar food abund-

	Cymatogaster aggregata	Embiotoca jacksoni	Hyperprosopon argenteum	Hypsurus caryi	Micrometrus minimus	Phanerodon furcatus	Rhacochilus toxotes	Rhacochilus vacca	Anisotremus davidsonii	Cheilotrema saturnum	Chromis punctipinnis	Girella nigricans	Hermosilla azurae	Medialuna californiensis	Oxyjulis californica	Paralabrax clathratus	Paralabrax maculatofasciatus	Paralabrax nebulifer	Semicossyphus pulchrum	Sarda chiliensis	Sebastes mystinus	Sebastes serranoides	Seriphus politus	Trachurus symmetricus	Atherinops affinis
Cymatogaster aggregata	X		X		J	X		X																	
Embiotoca jacksoni		X		X	J																				
Hyperprosopon argenteum	X		X		X	X																			
Hypsurus caryi		X		X		X	X	X																	
Micrometrus minimus			X		X																				
Phanerodon furcatus	X		X	X		X	X	X																	
Rhacochilus toxotes				X		X	X	X																	
Rhacochilus vacca	X			X		X	X	X	X																
Anisotremus davidsonii								X	X	X															
Cheilotrema saturnum									X	X															
Chromis punctipinnis											X				X						X	X	X		
Girella nigricans					X							X	X												
Hermosilla azurae												X	X												
Medialuna californiensis														X											
Oxyjulis californica											X				X										
Paralabrax clathratus																X									
Paralabrax maculatofasciatus																	X	X							
Paralabrax nebulifer																	X	X							
Semicossyphus pulchrum																			X						
Sarda chiliensis																				X					
Sebastes mystinus																					X	X			
Sebastes serranoides																					X	X			
Seriphus politus											X												X		
Trachurus symmetricus																								X	
Atherinops affinis																									X

J = juvenile

Table 6. Co-occurrence of King Harbor fishes in schools or aggregations.

ances have been noted by Straughan – (unpubl. data) – in the unstable soft-bottom substrate of King Harbor. Heterotypic schooling could be advantageous in this situation, increasing the visual field of the school. Heterotypic schools were often observed feeding, a situation not observed by Ehrlich & Ehrlich (1973) in Caribbean reef fishes.

We have only two examples from epibenthic species where mutual exclusion may result from food resource overlap. In both of these cases, water temperature serves as the primary isolating mechanism. The case for exclusion between *Hypsurus caryi* and *Embiotoca jacksoni* has been partially documented (Terry & Stephens 1976, Ellison et al. 1979). The other example involves three species of the serranid genus *Paralabrax*. These species are macropredators. *Paralabrax clathratus*, the kelp bass, a thermally tolerant species, usually lives in midwater under the kelp canopy, normally well above the rock substrate and above the thermocline. The other two species, *P. nebulifer*, the barred sand bass, and *P. maculatofasciatus*, the spotted sand bass, are benthic in orientation. The former prefers cooler, deeper waters of the outer coast while the latter is dominant in shallow embayments and prefers warm water. Separation of habitat position is distinct in the summer at King Harbor but tends to disappear in the winter when the inner harbor waters cool. At that time, all three species may be seen together, especially the two sand bass species. It would seem that some 'niche' separation is occurring here.

All in all, niche specialization or restriction (Roughgarden 1974) appears to contribute little to species packing in King Harbor. We also feel that unpredictability of the environment in the guise of predation sensu Paine (1966) and Connell (1975) is also relatively unimportant as inferred by low levels of observed predation and lack of habitat and nocturnal specializations, although fishing pressure may represent such an instability. In general, fishing pressure probably does not exceed that in other temperate metropolitan areas.

The diversity we observed appears to be the result of three primary factors: ecotonal nature of the environment (water temperature stratification); abundant food supply and high survivability of fish within the habitat; and regular fluctuations in reproductive success. The latter factor is a major contributor to non-equilibrium. The most 'stable' species were the resident viviparous embiotocids. The majority of the remaining common teleost species have pelagic eggs, embryos and larvae. Important exceptions include the ovoviviparous genus *Sebastes*, the demersal oviparous pomacentrids (*Chromis* and *Hyposypops*) and small benthic species (particularly blennies and gobies). There is an order of magnitude decrease in small benthic species between tropical and temperate reefs. Highly fecund species lacking parental care generally have erratic local success. At King Harbor, years of exceptional abundance have been observed for *Hypsypops*, *Chromis*, *Oxyjulis*, *Semicossyphus*, *Sebastes dalli* and *Lythrypnus dalli* since the beginning of this study. These temporarily expanded populations may then proceed to local extinction prior to a subseqent successful larval invasion (as seen in *Sebastes dalli* and *Lythrypnus dalli*). One of the broadcasting species which shows a stable pattern of abundance was *Paralabrax clathratus*, a species which apparently has an extremely short pelagic larval interval (Gardiner, pers. comm.). Table 7 presents some data on variations in the yearly abundance of juvenile fishes at King Harbor.

Table 7. Variations in abundance of juveniles by year (mean per occurrence/number of occurrences).

	Years					
Species	1974	1975	1976	1977	1978	Significant linear contrasts ($\alpha = 0.05$)
Atherinops affinis	0	118.7/7	749.0/2	999.0/1	128.3/8	74–5, 6; 75–7; 76–8
Chromis punctipinnis	56.8/18	68.3/35	162.7/14	78.7/14	44.9/18	74–6; 75–6; 76–8
Cymatogaster aggregata	90.8/10	106.1/22	230.1/6	0	0	74–6; 75–6
Embiotoca jacksoni	7.7/26	10.0/56	16.0/10	4.4/9	1.9/9	74–6; 75–8; 76–7, 8
Girella nigricans	18.0/9	6.8/16	25.1/6	23.7/4	12.5/2	75–6, 7
Hypsurus caryi	6.0/21	2.8/30	2.6/5	0	19.0/1	74–5, 7; 75–7; 76–7
Hypsypops rubicundus	2.3/18	2.5/65	2.6/20	2.8/30	2.8/8	
Oxyjulis californica	23.1/14	32.6/42	83.9/42	73.8/22	23.0/25	74–6, 7; 75–6, 7; 76–8; 77–8
Paralabrax clathratus	2.1/10	2.05/35	8.8/13	11.2/31	10.8/30	74–7, 8; 75–7, 8
Paralabrax nebulifer	0	0	1.5/2	4.2/10	5.2/13	
Phanerodon furcatus	3.2/6	3.1/16	1.0/1	0	0	
Semicossyphus pulchrum	0	1.0/3	0	3.4/19	3.4/10	
Rhacochilus toxotes	0	3.5/9	0	0	0	
Racochilus vacca	6.8/34	7.7/65	3.4/5	2.1/9	1.2/5	
Sebastes mystinus	65.4/26	22.1/35	5.7/12	30.5/4	0	
Sebastes serranoides	23.1/22	6.3/35	29.4/7	17.4/10	11.0/1	74–5, 6

Juveniles are often secretive and not accurately censused by transect methods. These observed variations, imply only relative differences not actual abundances. During this period, eggs, embryos and larvae sorted from meter net samples have also shown considerable variation (Jordan pers. comm.). We have considered the small number of species that produce demersal eggs in King Harbor may be related to the inability of warm temperate fishes to easily adapt their developmental processes to cooler water; reproduction being tied closely to spring and early summer periods when pelagic eggs, embryos and larvae can develop in warm nutrient-rich surface waters. By contrast, boreal cold-adapted fishes have larger, and usually demersal eggs (Marshall 1953). Barlow (1981) has suggested that the dominance of broadcasting on tropical atolls represents an adaptation for dispersal. We suggest that it probably represents the maintenance of the plesiomorphous condition of reproduction due to selective advantages in dispersal rather than representing a derived condition (apomorphous). We feel that invasion of the warm temperate zone may have been easier for species that have maintained pelagic developmental stages than those with demersal eggs, embryos and/or larvae). In any case, it is the large apparently random fluctuations in highly fecund species that result in much of the observed diversity. In this respect our results agree with those of Talbot et al. (1978) and Sale (1977) even though their work was in tropical habitat. A large portion of all marine fish diversity probably arises from the fact that they are non-equilibrium systems.

Acknowledgements

We wish to thank Byron Mechalas and John Palmer of Southern California Edison's Research and Development Division for their support. Karl Ehrlich and Joy Shrode of Occidental contributed laboratory data on thermal effects while Dale Straughan of the University of Southern California's Institute of Marine and Coastal Studies supplied data on food resources within the harbor. A large number of persons helped in the field, including Catherine Terry, John Ellison, Deane Terry, Steve Karsten, Pamela Morris, Steve Subber, Jay Carroll, Mickey Singer, Noriko Kawamoto, Kim Shriner. John Helly, University of California, Los Angeles, Dept. of Biostatistics, was responsible for data retrieval and analysis.

References cited

Barlow, W. 1981. Patterns of parental investment, dispersal and size among coral-reef fishes. Env. Biol. Fish. 6: 65–85.

Bray, R. N. & A. W. Ebeling. 1975. Food, activity, and habitat of three 'picker type' microcarnivorous fishes in the kelp forests off Santa Barbara, California. U.S. Fish. Bull. 73: 815–829.

Clarke, T. A. 1970. Territorial behaviors and population dynamics of a pomocentrid fish, the garibaldi, *Hypsypops rubicundus*. Ecol. Monogr. 40: 189–212.

Connell, J. H. 1975. Some mechanisms producing structure in natural communities: a model and evidence from field experiments. pp. 460–492. *In*: J. Cody & J. M. Diamond (ed.) Ecology and Evolution of Communities, Belknap Press, Harvard Univ., Cambridge.

Ebeling, A. W. & R. N. Bray. 1976. Day versus night activity of reef fishes in a kelp forest off Santa Barbara, California. U.S. Fish. Bull. 74: 703–717.

Ehrlich, K. F., J. M. Houd, G. Muszynski & G. E. McGowen. 1979. Thermal behavioral responses of selected California littoral fishes. U. S. Fish. Bull. 76: 837–849.

Ehrlich, K. F. , J. S. Stephens, Jr., G. Muszynski & J. M. Hood. 1979. Thermal behavioral responses of the speckled sanddab, *Citharichthys stigmaeus*: laboratory and field investigations. U. S. Fish. Bull. 76: 876–872.

Ehrlich, P. R. & A. H. Ehrlich. 1973. Coevolution: heterotypic schooling in Caribbean reef fishes. Amer. Nat. 107: 157–160.

Ellison, J. P., C. Terry & J. S. Stephens, Jr. 1979. Food resource utilization among five species of embiotocis at King Harbor, California, with preliminary estimates of caloric intake. Mar. Biol. 52: 161–169.

Houston, M. 1979. A general hypothesis of species diversity. Amer. Nat. 113: 81–101.

Love, M. & A. W. Ebeling. 1978. Food and habitat of three switch-feeding fishes in the kelp forests off Santa Barbara, California. U. S. Fish. Bull. 76: 257–271.

Marshall, N.B. 1953. Egg size in Arctic, Antarctic and deep-sea fishes. Evolution 7: 328–341.

Molles, M.C., Jr. 1978. Fish species diversity in model and natural reef patches: experimental insular biogeography. Ecol. Monogr. 48: 289–305.

Paine, R. T. 1966. Food web complexity and species diversity. Amer. Nat. 100: 65–75.

Poole, R. W. 1974. An introduction to quantitative ecology. McGraw-Hill, New York. 532 pp.

Roughgarden, J. 1974. Species packing and the competition function with illustrations from coral reef fishes. Theoret. Pop. Biol. 5: 163–186.

Sale, P. F. 1977. Maintenance of high diversity in coral reef fish communities. Amer. Nat. 111: 337–359.

Sale, P. F. & R. Dybdahl. 1975. Determinants of community structure for coral reef fishes in an experimental habitat. Ecology 56: 1343–1355.

Smith, C. L. & J. C. Tyler. 1972. Space resource sharing in a coral reef fish community. Bull. Natur. Hist. Mus., Los Angeles County 14: 125–190.

Stephens, J. S., Jr. & J. P. Ellison. 1977. A study of fish food habits as related to the biological enrichment of an area. Sea Grant-Fish Food Habits Symposium: 19–24.

Stephens, J. S., Jr. & J. B. Palmer. 1979. Can coastal power stations be designed to offset impacts by habitat enrichment? Mitigation Symposium, Fort Collins, Colorado: 446–450.

Talbot, F. H., B. C. Russell & G. R. V. Anderson. 1978. Coral reef communities: unstable, high-diversity systems? Ecol. Monogr. 48: 425–440.

Terry, C. B. & J. S. Stephens, Jr. 1976. A study of the orientation of selected embiotocid fishes to depth and shifting seasonal vertical temperature gradients. Bull. So. Calif. Acad. Sci. 75: 170–183.

Verner, J. 1977. On the adaptive significance of territoriality. Amer. Nat. 111: 769–775.

Abstracts

The foregoing papers are a selected, but hopefully representative sample from those presented at the second symposium on the Ethology and Behavioral Ecology of Fishes. Hopefully our selection will give a representation of the flavor as well as the content of the meeting. We have also included the Abstracts for all papers presented at the meeting, in alphabetic order. The papers were presented in a series of special sessions on particular topics, as outlined in the Preface, as well as general contributions. However, we have presented them here in alphabetic order for the convenience of readers interested in finding particular authors. In most cases the relation of the paper to particular sessions will be obvious, but in any event we felt that relationship was somewhat arbitrary, and gave little additional information.

We have exercised only minimal editing of Abstracts, other than sufficient to produce a uniform style for publication. Authors were allowed considerable leeway in length and content of Abstracts, and different amounts of time were allotted individual presentations in different sessions. In doing this we hope that the interest and excitement of ongoing research so evident at the symposium will be maintained. For it is this glimpse of ongoing science, and the reports of studies just completed or still in progress that have contributed to the success of these meetings.

In addition to those mentioned in the Preface, Moira Ferguson and Pat Noakes assisted in the preparation of these Abstracts for this publication.

The use of time and space by undisturbed wild brown trout (*Salmo trutta*) in a natural environment

Robert A. Bachman, *208 Mueller Laboratory, The Pennsylvania State University, University Park, Pa.*

Wild trout in their natural, undisturbed environment are shown to be more active, use cover less, and feed more frequently than those which have been tagged, handled, or constrained in so-called seminatural environments or laboratory conditions. It is suggested that stress of wild fish incurred as a result of handling and introduction to unfamiliar surroundings makes questionable the results of many previous studies which compared the behaviour of wild and hatchery fish.

The techniques and equipment used in the study of unconstrained, undisturbed wild brown trout in a limestone stream are presented. Included are methods of individual fish identification, length measurement by photographic techniques, and on site recording of agonistic behavior, feeding activity, and site utilization using a programmable microcomputer compatible with the Digital Equipment Corporation's System–10 hybrid computer. The use of core areas, home range, dominance hierarchies, type of feeding, feeding rates, and the mechanisms controlling population density and size structure are discussed.

The development of chemoreception in the young of the cichlid fish, *Cichlasoma citrinellum*

Colin Barnett, *Dept. of Zoology, University of California, Berkeley, Calif.*

At hatching *C. citrinellum* young demonstrate a chemical preference for their own species over blank water, but not for closely related species, who are in fact their major predators. The young are unable to distinguish between parental and nonparental fish. They respond most strongly to their own siblings, preferring them over any conspecifics. This preference for sibs increases with the number of sibs who are providing the chemical cue. These responses can be shown to be adaptive.

If the young are reared with only one parent, they prefer that parent over the other one. They show no evidence of a critical period for this preference.

On the first day of free-swimming only, young prefer the parent who spends the most time with them during the wriggler interval. This parent is invariably the female.

The cue for all chemical preferences is most likely the urine.

Behavioral response of yellowfin tuna, *Thunnus albacares* and kawakawa, *Euthynnus affinis*, to turbidity

Michael A. Barry, *National Marine Fisheries Service, NOAA and Dept. of Zoology, University of Hawaii, Honolulu, Hawaii.* Current address: *School of Life and Health Sciences, University of Delaware, Newark, Del.*

The effects of clouds of suspended sediments on the behavior of tuna were investigated as a part of the Deep Ocean Mining Environmental Study (DOMES). Captive yellowfin tuna and kawakawa were exposed to discrete clouds of suspended ocean mud from the DOMES region. Turbidity was measured in Nephelometric Turbidity Units (NTU). The behavioral responses of the tunas were recorded with a videotape system, and were quantified for statistical analysis. Both species were more sensitive in their detection and avoidance of turbidity than any other fish previously reported in the literature. However, the tuna suffered no obvious ill effects at the turbidity levels tested. Both kawakawa and yellowfin tuna significantly avoided turbidities of 1.7–1.8 NTU ($\sim$1.0 mg liter^1 DOMES mud), and detected turbidities as low as 1.4 NTU ($\sim$ 0.3–0.5. mg liter^1) relative to ambient levels of approximately 1.0 NTU. The response appears to be mediated by vision, and the tunas' ability to discriminate the clouds was reduced only on dark nights (0.3–0.5 lux). On brighter nights (3.0–8.8 lux), their discrimination appeared to be consistent with daylight results. The tuna would tolerate greater turbidities when motivated to feed in the clouds. The results are discussed in reference to ocean mining, tuna fisheries, and the behavioral ecology of tunas.

The evolution of agonistic behavior in amblyopsid fishes.

David L. Bechler, *Mundelein College, 6363 North Sheridan Road, Chicago, Ill.*

The Amblyopsidae, the cave, spring and swamp fishes, represent varying degrees of morphological, physiological and behavioral adaptation to a subterranean environment. With increased morphological and physiological adaptation to the cave environment, a concomitant decrease in overt agonistic behavior occurs. This is typified by reduced repertoire size, and temporal and spatial variation. Body size is a critical factor in determining the course and eventual outcome in bouts involving the least highly evolved subterranean species. The primary factor in determining the course and eventual outcome of bouts involving the more highly evolved subterranean species, appears to be territorial ownership, while body size is a secondary factor. The basic strategies employed by the different amblyopsid species during agonistic bouts suggest a conformation to Smith's Hawk-Dove and Hawk-Dove-Bourgeois models, thus providing a picture of the development of asymmetric factors involved in fighting.

The relation of habitat availability to social nesting for the longear sunfish (*Lepomis megalotis peltastes*) (Centrarchidae)

Brian F. Bietz, *Dept. of Zoology, University of Western Ontario, London, Ontario, Canada.*

Two forces were hypothesized to control the evolution of social nesting; restricted, patchily distributed nesting habitat, and social attraction between nesting conspecifics (which would arise if the proximity of a conspecific conferred a reproductive advantage). The nesting territories of male longear sunfish were found in close aggregations within a

400 m section of a small river in southwestern Ontario, and the effects of habitat availability on nest distribution were tested.

Using multivariate techniques (discriminant and cluster analysis), nesting habitat was delimited from other stream habitat. The multivariate description of nesting habitat was then used to predict the amount of available nesting habitat. Nesting territories were found to be aggregated within areas of predicted nesting habitat, and no biologically meaningful differences could be found between actual nest sites and those sites predicted as nesting habitat but which did not contain nests.

The available nesting habitat therefore appears underutilized and longear 'colonies' apparently occur not only as a result of specific habitat requirements but also from social attraction between males.

Migratory and territorial behavior of juvenile rainbow-steelhead trout (*Salmo gairdneri*) in experimental stream channels as related to reduced stream discharge: A progress report

Allen E. Bingham & Robert G. White, *Idaho Coop. Fish. Res. Unit, Fishery Resources, College of Forestry, Wildlife and Range Sciences, University of Idaho, Moscow, Idaho.*

Emigration of stocked wild and hatchery juvenile rainbow-steelhead trout (*Salmo gairdneri)* from experimental stream channels has been used to relate the effects of reduced stream discharge on stream rearing capacity. Experiments carried out in 1978 and 1979 indicated an increased tendency of the fish to emigrate from a reduced flow channel as compared to a constant flow channel.

The experiments have been conducted in two large flumes located on the Grande Ronde River near Troy, Oregon. The two identical flumes, 62.3 m in length and 6 m in width, have been partially filled with river gravel to stimulate a natural stream with a run-riffle configuration. Grande Ronde River water, diverted into a diked side channel, provided up to $0.57\,m^3\,s^{-1}$ in each channel. Upstream and downstream weirs and traps were used to monitor emigration of stocked fish and to prevent immigration of non-stocked fish.

The behavioral response of the fish to reduced stream discharge was observed by snorkeling and overhead visual observation in the fall of 1978 and summer 1979. Trout in the reduced flow channel exhibited a high degree of territoriality, while fish in the control channel exhibited virtually no territorial behavior.

Differential responses of obligate and facultative schooling juvenile fish swimming in currents

Eric C. Brinsfield & Melvin T. Huish, *Dept. of Zoology, North Carolina State University, Raleigh, N.C.*

Juveniles of two obligate schooling fish, *Brevoortia tyrannus* and *Mugil curema*, and a facultative schooling fish, *Leiostomus xanthurus*, were studied using 1114 individuals in a laboratory test flume. Swimming behavior was quantified by measuring the behavioral endurance time, which is the amount of time a fish, without extrinsic artificial stimulation, maintains its position in a current. Four treatments, consisting of groups of ten simultaneously swimming fish and solitary swimming fish in light and dark conditions, were studied by comparing the behavioral endurance times, with respect to circadian and lunar rhythms.

Analysis of variance indicated that all illuminated fish, except solitary *Mugil curema*, swam longer regardless of time of day. No statistical difference existed between groups and individuals in darkness, suggesting that grouped fish reap no energetic or orientation advantages over solitary swimming fish in the absence of light. In the presence of light, however, the mean response of groups was significantly greater than that of solitary fish for the obligate schooling species, but not for the facultative species.

The differences in behavioral endurance times of groups versus individuals of obligate schooling fish in light may provide an easily determined characteristic of schooling types. The basis of this difference, whether hydrodynamic or behavioral, is presently being analyzed through photographic measurements.

Buffers against aggression in *Betta splendens*

Paul M. Bronstein, *Trenton State College, Trenton, N.J.*

Male Siamese fighting fish (*B. splendens*) were studied to determine the behavioral mechanisms whereby aggression might be avoided. The following new findings were uncovered: First, when two animals were permitted to escape from the visual image of a conspecific partner, they typically choose *not* to aggress or display towards that other animal. Second, when two males meet in 20 liters of water, fighting is delayed for approximately 30 minutes while the animals alternately approach and then escape from each other.

Contrary to previous claims, the visual image seems neither simply attractive for *Bettas*, nor is that stimulus a positive reinforcement for approach responses. The visual image of a *Betta* elicits both approach and withdrawal movements. Fighting occurs only after the persistent approaches of an opponent overwhelm another animal's escape maneuvers. Fighting is much reduced in large, complex tanks which permit fuller expression of escape responses.

The comparative behavioral ecology of two species of co-inhabiting tide pool blennies

G. Thomas Chandler & David G. Lindquist, *Dept. of Biology, University of North Carolina at Wilmington, Wilmington, N.C.*

The Feather Blenny, *Hypsoblennius hentz*, and the Crested Blenny, *Hypleurochilus geminatus* are co-inhabitants of rocky tidepools at Fort Fisher in southeastern North Carolina. Because both species coexist in close geographic association in this limited, intertidal reef habitat, a natural laboratory was provided for an intergeneric investigation of the comparative behavioral ecology of these two phylogenetically related organisms.

We attempted to determine whether these two species are competing ecologically or, if not, how competition is avoided. Five major areas of comparative ecology were researched: reproductive behavior, population structure, feeding behavior, habitat selection and agonistic behavior.

Hypsoblennius hentz males construct nests most frequently inside abandoned oyster and whelk shells, whereas *H. geminatus* nests are usually found flush against a hard substrate (rock or piling) hidden by algae and/or hydroids. During the reproductive months, *H. geminatus* males develop hooked shaped tabs subdistally to the tips of the anal rays and probably use them in nest tending.

Gonad indices of male and female *H. hentz* indicate a peak breeding season in June, July and August, whereas gonad indices for *H. geminatus* indicate a phase shifted breeding season peaking in late July, August and September.

In both species, the relative abundance and size of individuals increase from March to June, but during the aggressive reproductive months fewer and larger individuals occur. A marked increase in abundance and decrease in size is characteristic of both populations in October. In November, as temperature and food resources decline, the populations emigrate possibly to subtidal waters.

The diets of both species are almost identical and consist primarily of caprellid and gammarid crustacea, followed by algae, hydroids and polychaetes. Smaller fish (15–45 mm S.L.) selectively feed most frequently on crustacea, whereas larger fish (45–75 mm S.L.) tend to graze over the substrate and consume crustacea and significantly larger amounts of algae.

Habitat selection studies indicate that habitat matching occurs in both species with smaller fish occupying small diameter refuges and larger fish occupying large diameter refuges. Habitat matching serves as an effective anti-predator–anti-aggressor mechanism in both species.

Agonistic behavior in *H. geminatus* is characterized by a mottled saddle pigment pattern over the body and a dark head pattern with two contrasting diagonal post orbital bars. The agonistic motor patterns in order of increasing aggressive intensity are: median fins erect, head up, gular puff, approach, quiver, tail beat, gape, bite and chase. Head down, body tilt, median fins depressed and flee are characteristically submissive postures. The two lower, enlarged and recurved dentary canine teeth are used by the aggressor to scrape or impale an opponent. Agonistic behavior is currently being investigated for *H. hentz* along with intergeneric influences on habitat selection and agonism.

Social spacing in the blackeye goby.
K.S. Cole, *Dept. of Zoology, University of Alberta, Edmonton, Alberta, Canada.*

Coryphopterus nicholsi, the blackeye goby lives subtidally to a depth of 25 m in rock-rubble zones. Its range extends from Baja California to the Queen Charlotte Islands of British Columbia, and it is common in its usual habitat. Both sexes are territorial throughout the year, although only the male is involved in parental care. Field observations and laboratory experiments indicate that defense of cover for protection from predators is a likely benefit of territoriality. Laboratory experiments currently are in progress to compare the use of space by each sex, and by various age groups, during both the reproductive and non-reproductive seasons. Preliminary results of these laboratory experiments will be discussed. In particular, changes in area and amount of cover defended will be compared for non-reproductive, courting, and parental males.

Movements of *Micropterus salmoides* in relation to the water quality of the Atchafalaya Basin, Louisiana
J.F. Doerzbacher & C.F. Bryan, *Louisiana Coop. Fish. Res. Unit, 245 Ag Center, Louisiana State University, Baton Rouge, La.*

The Atchafalaya Basin, the largest hardwood river swamp in North America, occupies about 4680 km^2 in the leveed floodplain of the Atchafalaya River. Depletions of dissolved oxygen (DO) are commonplace in the lower portion of the Basin. Radiotelemetric techniques were used for tracking fishes in the swamp, bayou, and shallow lake habitats to study movements and behavior of fishes in response to a continually changing hydrology and water quality in selected habitats of the lower Basin. Transmitters, approximately 30 mm long, 5 g in weight, and with a life expectancy of 3 months were surgically implanted while in the field and fish were released immediately. Preliminary results from 8 bass tagged beginning in July 1978 and tracked up to 85 days indicate that it is questionable whether movements were directed by changes in DO. However, no serious condition of low DO was encountered while tracking these fish. Tagged fishes moved from a water column comparatively lower in average DO to one higher in DO on 20 of 39 observed changes in location. Tagged fishes moved on 22 of 39 occasions to a water column with an average temperature closer to a 'preferred' temperature derived from the literature. An analysis of movement in relation to pH, specific conductance, and oxidation-reduction potential has not been completed.

Female choice and the evolution of sexual dimorphism in the mottled sculpin
J.F. Downhower, R. Pederson & G. Staples, *Dept. of Zoology, The Ohio State University, Columbus, Ohio*, & L. Brown, *Dept. of Biology, George Mason University, Fairfax, Va.*

Male mottled sculpins are larger than females (ca. 25%), are polygynous and undertake parental care duties during the post-spawning season. Females show preferences for large males. Large males are the first to obtain spawnings although many smaller males will obtain spawnings later in the breeding season.

Female preference for large males has two direct effects. Larger males remain longer at their spawning sites, and egg losses during the spawning season are lower at sites occupied by larger males. Males do not construct their breeding sites which are located under rocks or other streambed rubble. Males can only feed opportunistically during this period, hence large size may directly influence the ability of a male to remain at a spawning site.

Male parental care activities concern defense of the eggs from predators. The chief predators on sculpin eggs appear to be other sculpins – particularly spawned females. Predation by females insures that females will show preferences for males who are larger than themselves since aggressive encounters are size-mediated. Thus intersexual aggression assures that males are the larger sex.

Sculpins are cannibalistic when and if the opportunity arises. During the spawning season large males may ingest small females. Small females generally spawn at high risk since they are the last to spawn and therefore do so when males have undergone a prolonged period of fasting.

The interaction between egg predation by females and cannibalism by males sets both the lower and upper limits of sexual dimorphism in this species.

Ontogeny of diel rhythms of swimming activity and of vertical distribution in juvenile pink salmon (*Oncorhynchus gorbuscha*)

Jean-Guy J. Godin, *Dept. of Zoology, University of Western Ontario, London, Ontario, Canada.*

In late summer or fall adult pink salmon spawn in streams and rivers of the west coast of North America. The female buries her eggs in a nest (redd) in the gravel substrate. The following spring young fry (28–35 mm) emerge from the redd, and shortly afterward migrate downstream, mainly at night, to the ocean. The fry are pelagic, diurnal feeders while residing in estuaries and adjacent coastal waters. Little is known about their diel (daily) pattern of emergence from gravel and their subsequent diel activity rhythms.

In this paper I describe 1) the diel timing of fry emergence from a simulated gravel redd at different water temperatures, and 2) the ontogenetic changes in their diel rhythms of swimming activity and of vertical distribution after emergence.

Each of six simulated redds used had a different mean temperature (range: 3.4–15.0° C). All redds were exposed to a light-dark cycle of LD 12:12. Newly hatched sibling embryos were introduced in each redd, and allowed to emerge spontaneously. Emerging fry were collected in a container at the outflow of the redd. The container was emptied periodically every day and emerged fry counted.

Diel rhythms of swimming activity and of vertical distribution in groups of six fry each were recorded automatically with the aid of a video camera and recorder for 37 days after emergence. Each of six different groups of fish were observed independently and sequentially for 5-day periods in an aquarium exposed to overhead illumination (LD 12:12) and 10° C.

Results showed that emergence from gravel was mainly nocturnal at five of six temperatures. No distinct diel pattern was observed at 15° C. The diel pattern of emergence was synchronized with the onset of darkness, but temperature affected the timing of this rhythm. The rate of sibling fry emergence for the entire period of emergence increased with decreasing redd temperature.

Fry were night-active during the first week after emergence. During this period the fry were also swimming closer to the water surface at night and closer to the tank bottom during the day. The fry gradually became more day-active (diurnal) between Days 7 and 10 after emergence. Thereafter until the end of the experiment, distinct diurnal activity rhythms were observed. Coincident with this ontogenetic shift in the diel rhythm of swimming activity was an increasing tendency of the fish to swim near the water surface during the day after Day 8. This observation suggests that the negative photoactic response of newly emerged fry weakens gradually with age. However, the fry always swam closer to the water surface during the night than during the day throughout the experimental period. Diel rhythms of swimming activity and of vertical distribution were strongly synchronized with the LD cycle.

Nocturnal emergence is considered an anti-predator adaptation, and a positive photoresponse and a diurnal swimming activity rhythm are considered adaptive for schooling and feeding during the day.

Territoriality and sexual selection in the johnny darter (*Etheostoma nigrum*)

J.W. Grant & P.W. Colgan, *Dept. of Biology, Queen's University, Kingston, Canada.*

A field study was undertaken to determine the function of territoriality in the reproductive behavior of the johnny darter. Male johnny darters defend a territory around a rock nest from a few days prior to spawning until their eggs hatch. Females leave the nest immediately after spawning.

The frequency of territorial behaviours performed by nesting males was initially high and decreased over the reproductive cycle. Similarly the distance that intruders were chased decreased over the reproductive cycle.

Although there are undoubtedly nest site and egg defense functions to territoriality, we suggest that the conspicuous nature of the behaviour serves a sexual selection function. Preliminary results indicated that the best predictor of a male's reproductive success was his mean distance of response (MDR) to females. Males that travelled farther from their nesting rock to court or respond to females acquired more eggs. MDR to females correlated positively with MDR to other intruders, and a male's overall MDR was a good indicator of his ability to raise and defend his eggs. Thus females may be using MDR as a cue in mate selection.

Feeding behavior of territorial and nonterritorial cunners, *Tautogolabrus adspersus*, in Conception Bay, Newfoundland

John M. Green & William David Martin, *Dept. of Biology and Marine Sciences Research Laboratory, Memorial University, St. John's, Newfoundland, Canada.*

The feeding behavior of territorial and nonterritorial cunners, *Tautogolabrus adspersus*, in Conception Bay, Newfoundland is described. Both behavioral types feed primarily on benthic food items. Significant differences exist, however, between the temporal aspects of the feeding behavior of territorial and nonterritorial fish, and the diets of territorial males, nonterritorial males and nonterritorial females significantly differ from each other. These differences are discussed in relation to the function of territoriality in this species.

Post-spawning behavior of male lumpfish, *Cyclopterus lumpus*, Cyclopteridae, in Conception Bay, Newfoudland

John M. Green & Tim Shears, *Dept. of Biology and Marine Sciences Research Laboratory, Memorial University, St. John's, Newfoundland, Canada.*

Post-spawning male lumpfish, *Cyclopterus lumpus*, were observed in Conception Bay, Newfoundland in order to describe and quantify their behavior. Males aerate the eggs both by pectoral and caudal fanning and by 'blowing' water through them. Slowly moving predators such as sea urchins and sea stars are carried away from the egg mass, while more active egg predators such as other fish are chased.

Males spend variable lengths of time away from their egg mass both daily and over the incubation period. Variation among fish in the time spent away from the egg mass appears to be related to the characteristics of the nest site, and is related to the hatching success of the eggs. Quantitative data on fanning and predator removal are discussed in relation to the hatching success of individual egg masses.

Sexual selection in sticklebacks: resource prediction

Harm P. Gross* & Dierk Frank, *Universität Hamburg, Zoologisches Institut und Zoologisches Museum, 2000 Hamburg 13, F.R.G.*

* Present address: *Dept. of Biology, Queens' University, Kingston, Ontario, Canada.*

Sexual differences in reproductive strategy have been attributed to relative parental investment (Trivers, 1972). In general, males appear to invest less than do females who develop energetically expensive eggs and may also care for the offspring. Sexual selection in such animals consists of epigamic selection (choice) by females and intrasexual competition by males. Choosers are expected to decide among perspective mates on the basis of predicted future resources for their offspring (Williams, 1975). The operation of sexual selection in animals with partially reversed strategies is poorly understood. This study reports on sexual selection in one such species.

Reversal in the sex providing parental care occurs in sticklebacks as offspring are reared by solitary males. Secondary sexual characteristics are not reversed, however. Experimental trials demonstrate that reproductive success of male three-spined sticklebacks (*Gasterosteus aculeatus*) varies with levels of courtship activity and agonistic behavior towards neighbouring males (which also determines territory size). We interpret these relationships as follows. Male success in obtaining females increases with courtship activity because certain behavioral elements of this activity (especially the first, zigzagging) are used by males to advertise, and by females to predict, future paternal resources for the offspring. We believe that chief among these resources is protection of developing eggs and fry from cannibalistic neighbouring males, schooling females, and other predators. This protection (and hence reproductive success for both males and females) will increase with agonistic behavior by males following the completion of courtship. Females can estimate expected levels of male aggression from male courtship since the intensities of these behaviors are correlated among males. It is argued that cheating by males during courtship could not evolve in sticklebacks.

Field studies emphasize the importance of male reproductive behavior relative to nest-site characteristics in determining reproductive success.

We suggest a genetic mechanism for inter-male differences in courtship behavior, and an explanation for the origin of male territoriality and paternal care in sticklebacks.

Faces in the crowd: School fidelity and individuality in yellow perch

Gene S. Helfman, *Dept. of Zoology and Institute of Ecology, University of Georgia, Athens, Ga.*

It is common for a school of diurnally foraging fishes to be sighted repeatedly in the same area of a lake or reef for periods of days or even years. An answer was sought to the question of whether such schools are (A) cohesive, extended-tenure units made up of the same individuals over long periods; (B) the shorter-term result of overlapping individual home ranges of fish with foraging requirements best served by a schooling strategy; or (C) convenient assemblages composed of nomadic individuals with varying schooling tendencies that break up and reform as immediate responses to environmental factors.

The question was approached by marking 200 schooling and non-schooling yellow perch (*Perca flavescens*) with individually recognizable tags. Diving observations of these fish were made in a 1.3. ha bay in Cazenovia Lake, N.Y. over a two-year period. Location within the bay, school composition, school size, and individual behavior were recorded. The answer to the question of school fidelity and personality is (D) all of the above, depending on the individual fish and time interval involved.

The role of social interactions in toxicity testing

Mary G. Henry & Gary J. Atchison, *Dept. of Animal Ecology, Iowa State University, Ames, Iowa.*

The numerous and significant technological advances of the last ten years have brought with them many environmental problems that are just as significant. It is, therefore, crucial that we develop new methods to evaluate the impact of pollutants on life forms that are rapid, sensitive and biologically meaningful.

It was the aim of this research to develop and utilize a behavioral bioassay to assess the effects of sublethal levels of cadmium and zinc on bluegill, *Lepomis macrochirus*. One of the most important aspects of this study was to examine the effect of the metals on social order and behavior, since many species of fish, including bluegill, are known to form social groups in nature. Social order and social groups have evolved for a reason, whether to increase mating opportunities, increase feeding efficiency, improve predator defense, etc. Consequently, consideration of an animal's social system is important in obtaining information regarding its survival in a contaminated ecosystem, since group living helps insure reproduction and survival, thereby increasing fitness of individuals and groups.

Disturbances due to pollutants may change the competitive advantages possessed by some fish. The results of this study indicate that levels of cadmium and zinc as low as 21 $\mu g/l$ and 99 $\mu g/l$, respectively, can significantly and differentially alter the frequency of performed comfort movements and aggressive interactions in individuals of different social rank.

This paper will compare the literature to behavioral changes occurring in bluegill populations exposed to five sublethal levels of cadmium and zinc. Hopefully, these findings will encourage research in the area of toxicity testing and point out the need for the advantages in using behavior as the final integrated measurement of the impact of a toxicant on the biochemical and physiological processes of fish in a contaminated environment.

Squid schools: Are squid really just invertebrate fish?
Ann C. Hurley, *Moss Landing Marine Laboratories, P.O. Box 223, Moss Landing.*

Although studies of schooling have concentrated on the behavior of fishes, there are many invertebrates which school as well. A comparative approach which examines the similarities and differences between some of these invertebrate schools and fish schools may give clues as to the types of evolutionary pressures which shaped schooling behavior. It is likely that many squids, in particular, have been subjected to the same environmental constraints as many of the obligate schooling fishes. On the other hand, their mode of locomotion and general physiology are quite different. This paper will examine aspects of the schooling behavior of the squid *Loligo opalescens* and will compare what is known about the details of school structure in squid to studies done with fish.

The interactions of squaretails (*Tetragonurus* spp.) with pelagic tunicates, particularly salps
John Janssen, *Dept. of Biology, Loyola University, 6525 North Sheridan, Chicago, Ill.*

Tetragonurus is a rarely collected stromateoid of the open ocean from tropical to temperate regions. Observations while scuba diving indicate the young live inside of salps and all three species (*T. cuvieri, T. atlanticus,* and *T. pacificus)* have been collected from them. Stomach analysis indicates that they feed primarily on pieces of salp and pyrosome that they bite off by means of an unusual dentition. Observations of fish in aquaria indicate the fish enter salps via the atrial opening. While inside they may eat the salp's gut and gill bar. If the salp is small the gut is eaten from the outside by chewing through the salp's test. Very small salps can be consumed whole. The salps are not without defenses: they can prevent entry into the branchial chamber by tightly closing the atrial opening; a tough 'shield' around the gut, test processes, and the arrangement of individuals in a chain inhibit attacks from the outside. The fish also consume test processes of *Pyrosoma* and enter its cloacal chamber. Individual zooids are seldom eaten as the test processes protect them. Salps with guts eaten out are regularly observed while diving.

A number of morphological characters suggest that *Tetragonurus* is specifically adapted for living in tunicates and using them as food. In turn, salps appear adapted to resist *Tetragonurus* and predators that feed like it. The present study and other field observations indicate that associations like this may be common in the open ocean.

Behavioral mechanisms of reproductive isolation in *Etheostoma spectabile* (Agassiz) and *Etheostoma caeruleum* (Storer)
Lin. J. Kozlowski, *Box 27, Williamsburg, Miss.*

This study was conducted to observe, describe, analyze, and compare the reproductive behavior of *E. spectabile* and *E. caeruleum* in order to determine what behavioral mechanisms help maintain reproductive isolation in these two species in South Central Missouri. The general behavior of *E. spectabile* and *E. caeruleum* is also described and compared.

An artificial stream was constructed that held approximately 50–60 gallons of water and was maintained on a 13.5 hour daily light regime. Ten darters were housed in this environment and observed for two months. For the next two months observations took place in the natural streams in South Central Missouri during the breeding season.

Although the darters in the artificial stream system increased in color intensity, they never reached the maximum color intensity of the darters in the field. Courtship and spawning behaviors were not observed in the laboratory.

Slight differences do occur in the ecological habitat and in the time of the breeding season, but both species occur in the same habitats and the spawning time does overlap. Very little species interaction indicated that these two species remain distinct in South Central Missouri. There is a slight behavioral difference in the courtship pattern in which *E. spectabile* utilizes Head Wagging as part of the sequence, and *E. caeruleum* does not. Sex recognition may also take place by the males.

In conclusion, this preliminary study indicates that *E. spectabile* and *E. caeruleum* remain distinct species by habitat isolation, seasonal isolation, and behavioral isolation. Although each of these is a slight or fine difference, each contribute to the maintenance of distinct species.

Effects of schooling prey on the hunting behavior of a fish predator

Terrance M. Lim, *Dept. of Zoology, University of California, Berkeley, Calif.*

Lone *Brachydanio rerio* and schools of 2, 8, 16, 24, and 32 were presented to single *Haplochromis polystigma* in order to study the effects of the different levels of prey availability on hunting efficiency and predatory behavior. Ten experienced predators were used and each was presented with all six levels of prey number. The sequences of behaviors in all the resulting predator-prey interactions were recorded according to categories established in earlier experiments. Measures were taken to control for the effects of satiety and to assure that prey were uniformly susceptible to the predator in any given experiment.

The proportion of successful attempts at capturing prey (hunting efficiency) decreased as the number of prey increased. The best explanation for this trend appears, as suggested previously, to be that of a confusion effect. This hypothesis will be examined by comparing the transition frequencies of behaviors in the predators' hunting sequences between prey levels. It appears, as would be expected, that both simple arithmetic and perceptual processing phenomena are acting in favor of schooling prey.

The effects of temperature on the ability of bluegills to avoid sublethal gradients of ammonium chloride.

Kenneth S. Lubinski, *Illinois Natural History Survey, Havana, Ill.*

Avoidance responses of individual bluegills, *Lepomis macrochirus* (Raf.), produced by abrupt, sublethal gradients of ammonium chloride (approximately 40 $\mu g/l$ un-ionized ammonia) were monitored at 14 and 22 C. All fish were exposed in a square laminar flow tank in which a median longitudinal gradient was produced. The fish tested at 14 C were monitored during a 1-hr control period followed by a 1-hr test period in which the ammonium chloride was pumped into one side of the tank. The control and test periods were extended to 5 hr each for the fish tested at 22 C due to the use of a television-computer monitor which tracked the fish automatically for extended periods of time.

The bluegills exposed at 14 C initially exhibited decreased activity and visible stress but failed to avoid the ammonium chloride zone even after becoming aware of the uncontaminated area. The majority of the fish tested at 22 C did avoid the ammonium chloride zone and none were visibly stressed. These results indicated that low temperatures may increase the sensitivity of bluegills to ammonia in spite of the fact that the proportion of the toxic chemical form of ammonia, un-ionized ammonia, is reduced at low temperatures. The implication of these test results is that fish populations in stream reaches below sources of ammonia, such as sewage treatment plants and certain industries, may be more susceptible to such effluents in winter than in other seasons, a situation which is generally compounded by the fact that ammonia is converted to less toxic substances more slowly at low temperatures.

Seasonal variation in abundance of three behavioral classes of sailfin molly males, *Poecilia latipinna*

Carol L. Luckner & J. Michael Fitzsimons, *Dept. of Zoology and Physiology, Louisiana State University, Baton Rouge, La.*

Males of the Sailfin molly, *Poecilia latipinna*, are highly variable in both morphology and reproductive behavior. Three classes of males can be defined on the basis of the origin and insertion of the dorsal fin, proportional measurements of dorsal-fin basal and depressed lengths, pre-dorsal length, caudal peduncle depth, body depth, and caudal-fin length. These same three classes of males can also be defined on the basis of their reproductive behavior: males exhibiting no displays, dominant males who consistently display, and males who, depending on their dominance status, may or may not display. All classes of males often occur synchronously in the same population but each is more prevalent at certain times of the year. Peaks in numbers of large males and of small males are asynchronous. The behavioral and ecological significance of the male classes is discussed.

Social dispersion in northern gulf killifishes

Lorraine Madsen & John Michael Fitzsimons, *Louisiana State University Museum of Zoology, Baton Rouge, La.*

Social dispersion patterns were studied in the killifishes *Adinia xenica, Cyprinodon variegatus, Fundulus grandis* and *F. similis* from syntopic populations on the Louisiana Gulf coast during field and laboratory observation. Behavioral units of spacing and grouping activity were defined, classified and named. Fishes were observed to determine how these behaviors were affected by phylogenetic relationship, morphology, and habitat type. Strikingly similar agonistic behavioral sequences (spacing behavior) were composed of 11 aggressive and aggression-reducing components of which nine were common to all four species. Similarity of agonistic behaviors were likely due to convergence of components associated with breeding territoriality to facilitate recognition of signal functions between related, syntopic species. Similarity of body form had more influence on schooling tendencies (grouping behavior) than reported phylogenetic relationship. Schooling tendencies were significantly higher among conspecifics, within species with only slight sexual dimorphism, and between species with similar morphotypes. School structures varied with changes in habitat. In quiet pools, schools were small, spherical, often heterotypic, and slow-moving; whereas, in narrow waterways, schools were large, elongate, usually homotypic, fast, and dense. The social organization of these killifishes, dependent upon environmental conditions, is expressed in dispersion patterns which include territorial males evenly spaced during spawning, randomly dispersed individuals feeding among vegetation during high water, and conspecific and heterospecific schools of differing structures.

A comparison of the spawning behavior of hormone sex-reversed *Tilapia* and untreated fish

Frank H. Meriwether & William L. Shelton, *Ethology and Behavioral Ecology of Fishes, Auburn University, Auburn, Ala.*

Various species of the family Cichlidae are important food fishes throughout the world but excessive reproduction adversely affects the entire population. Methods to control unwanted reproduction have included monosexing by manual separation, interspecific hybridization, or hormone sex reversal. The sex reversal technique produces reproductively functional individuals, one-half of which have the same sexual genotype and phenotype while the other one-half have the phenotype appropriate to the hormone treatment but with the opposite genotype. These two groups are identified by individual pair spawnings in aquaria and sexing the resulting progeny. Spawning behavior has been observed on sex reversed females (genotypic males) of *Tilapia aurea* and concurrently on untreated females (genotypic females) each mated with normal males (genotypic males).

Space utilization and movement of adult largemouth bass in a small southeastern reservoir

George Clark Mitchell, *Southeast Georgia Regional Office, Environmental Protection Division, Brunswick, Ga.* & Carl Quertermus, *Biology Dept., West Georgia College, Carrollton, Ga.*

Fifteen largemouth bass, *Micropterus salmoides*, were tracked in Lake Carroll, Carroll County, Georgia, using ultrasonic telemetry techniques. Recorded data were used to determine space utilization and day and night movement. Lake Carroll largemouth bass seem to be of two behaviorally different types, those that utilize a small, well-defined home range, and those that utilize large portions of the lake. Of the 15 bass tracked, eight occupied home ranges, six used large portions of the lake, and no conclusions could be made on the other. For all seasons studied (spring, summer, and fall), the majority of fish moved more during the day than at night. A paired comparison for all fish showed that day movement was significantly greater than night movement.

Some effects of prolactin on the parental behavior of the convict cichlid, *Cichlasoma nigrofasciatum* (Gunther)

Robert E. Muller, *3 Roosevelt Road, Springfield, Ill.*

Some experimental evidence shows that hormones play a role in fish behavior. The experiments carried out in this study have contributed to that conclusion. In the first experiment, convict cichlids were injected with ovine prolactin (Ringer's solution in control fish) following spawning. Artificial eggs were substituted for the real eggs and the amount of egg fanning behavior was measured. In those fish receiving the prolactin injections, the intensity of fanning (percent fanning per trial) of experimental fish exceeded that of control fish. Also, the number of days that the experimental fish continued to fan the artificial eggs far exceeded the controls.

In another experiment, non-breeding female convict cichlids were injected with prolactin and then presented with a batch of newly laid eggs. In most cases the prolactin injected fish did not eat the eggs whereas in all cases the eggs were eaten by control fish.

Following spawning the amount of pit digging behavior was measured in fish receiving prolactin injections and control fish. The control fish had a higher level of pit digging behavior than the experimental fish indicating that prolactin inhibits this behavior.

Inexperienced, non-breeding convict cichlids were injected with prolactin and presented with conspecific fry. However, the fry were eaten and the data failed to show that prolactin injections could induce egg eating inhibition and parental care.

Convict cichlid pituitary glands were removed for study at three different times: in non-breeding fish, in fish engaged in egg-fanning, and in fish during care of free-swimming fry. The prolactin secreting eta cell showed a greater level of activity in the breeding fish with the greatest level of activity in the fish which were fanning eggs.

Diversity of spawning behavior in darters (Percidae)

Lawrence M. Page, *Illinois Natural History Survey, Urbana, Ill.*

Among species of darters for which the egg-deposition site is known (39 of about 148 species, representing 17 of 28 subgenera), two general modes of spawning behavior exist. One is to bury the eggs in the substrate. The other is to attach the eggs to an object, usually a rock or plant, above the substrate. Egg-burying is characteristic of primitive species, including all species of *Percina* for which data are available, and of some of the more highly evolved species, including species of the subgenera *Oligocephalus* and *Nothonotus* (of *Etheostoma*). Egg-attachment, a derived behavior, is known only in *Etheostoma*, but has arisen independently in several subgenera. The development of egg-attachment has been accompanied by functional adaptations of the female genital papilla.

During spawning an egg-burying female works her body partially below the surface of the substrate and with her genital papilla buried expels eggs. The substrates usually utilized are loose gravel, sand, or mixed gravel and sand. The papillae of egg-buriers are conical tubes.

Two groups of species may be recognized among egg-attaching species. Those which assume an inverted position and arrange their eggs in a cluster on the undersides of stones (*E. (Nothonotus) maculatum* and species of the subgenera *Boleosoma* and *Catonotus* (of *Etheostoma*) have wide, flat papillae. Those which assume a vertical or horizontal (including sometimes inverted) position and attach eggs to plants, stones, etc., exhibit a diverse array of papillae.

Sensory factors and the 3-dimensional structure of fish schools

Brian L. Partridge, *Dept. of Biology, University of Miami, Coral Gables, Fla.*

A number of studies have investigated aspects of the sensory basis of fish schools and their internal structure or have speculated about the function (survival value) of schooling. Little attempt has been made, however, to integrate these topics. In this paper I will try to tie together results from a number of recent studies to show that structure of schools results from the constraints posed by the sensory capabilities of the fish and, ultimately, from the function of schooling as an antipredator device. The work I will discuss was carried out at the Department of Agriculture and Fisheries for Scotland Marine Laboratory, Aberdeen and for the most part was undertaken jointly with T.J. Pitcher.

In several papers resulting from that work, we have shown that 1) fish schools *do* have recognizable (i.e. non-random) structure. Surprisingly, many aspects of internal structure are constant across species as dissimilar as saithe, cod, herring and European minnows. In contrast to some previous suggestions that structure depends upon absolute size of the fish it now appears that differences in structure between these species are best explained in terms of differences in maneuverability and turning radius of the fish. Altering sensory input by lateralis-section or temporary blindfolding individual fish within normal schools or entire schools of saithe causes large changes in the positions fish take up with respect to their neighbors. Separate experiments on the sensory capabilities of the fish demonstrate that the changes exhibited by sensorily deprived fish result in their taking up those positions with respect to neighbors at which they can best determine their headings and velocities by vision or lateral line alone. School structure is thus seen to result from the sensory capabilities of the individuals within it. Experimental evidence does not support the suggestion that structure results either from fish packing into the smallest possible space or from the hydrodynamics of swimming. Fish take up those positions at which information is most rapidly transferred between individuals and collective decision making is enhanced.

Behavioral ecology, the key to environmental assessments of stream habitat

C.G. Prewitt & C.B. Stalnaker, *Coop. Instream Flow Service Group, US Fish and Wildlife Service, Fort Collins, Colo.*

Recent emphasis upon responses of aquatic organisms to environmental change has led to related activities in diverse fields of environmental science. The basis for most assessment ability within these fields has involved use of physical (streamflow) temperature and chemical (water quality) predictive models associated with known response characteristics of fish species of concern.

While physical properties may be accurately predicted in most situations, the reaction of a given species to changes in these properties is often a composite of physiologic and behavioral mechanisms. As a result, it is often difficult to fully predict changes in stream fish populations or community characteristics because we know little of the role of behavioral dynamics in determining population or community attributes.

Studies in progress at the US Fish and Wildlife Service, Cooperative Instream Flow Service Group (CIFSG), lend support to theories relating population and community character states to physical stream conditions. Within simple communities, it appears possible to predict population attributes (standing crop, biomass, etc.) resulting from alterations in flow or water quality.

Effects of behavioral mechanisms in more complex communities and the nature of their alterations under altered conditions are largely unknown.

Field-oriented experiments which document behavior-related phenomena under a variety of instream conditions are needed as supporting concepts in environmental assessment programs.

The authors stress quantified results and statistical supportive designs and offer suggestions for increased interaction among researchers in the physical and behavioral disciplines. Initial data on behavioral criteria of selected fish species and life stages are presented.

Habitat selection of adult largemouth bass in a small southeastern reservoir

Carl Quertermus, *Biology Dept., West Georgia College, Carrollton, Ga.*, & George Clark Mitchell, *Southeast Georgia Regional Office, Environmental Protection Division, Brunswick, Ga.*

Fifteen largemouth bass, *Micropterus salmoides*, were tracked in Lake Carroll, Carroll County, Georgia, using ultrasonic telemetry techniques.

Three aspects of fish habitat selection (distance from shore, water depth and dock utilization) were analyzed in relation to selected variables. The depth of water selected was significantly related to water temperature and water clarity. Fish were found in shallower water at lower water temperatures and reduced water clarity. The only variable significantly related to dock utilization was barometric pressure. When fish were in water less than 1.5 meters they utilized docks more often at higher barometric pressures. No significant relationships between fish distance from shore and the independent variables were found. Analysis of habitat utilization by season (spring, summer, fall) showed fish (1) were closer to the shore in spring than either summer or fall, (2) stayed in shallower water in spring and fall, and (3) utilized docks more in fall than either spring or summer.

Schooling as a means of enhancing the ability of fish to respond to environmental stimuli

Scott L. Ralston, *Dept. of Biological Sciences, University of Southern California, Los Angeles, Calif.*

Fish in a school show enhanced ability to respond to laboratory stimuli. This enhancement is roughly proportional to the school size, up to a critical number of individuals. Groups of top smelt, *Atherinops affinis*, of from one to fifty individuals were tested with laboratory experiments including natural fright response threshold determinations with variable light CS and electric shock UCS, and shuttle box shock avoidance conditioning with variable light CS and electric shock UCS. This enhanced ability to respond was demonstrated to be the result of three factors: 1) greater directional coverage by larger schools, 2) increased response consistency near perception thresholds with larger schools, and 3) an actual lowering of the response threshold with larger schools up to the critical limit. This critical number for top smelt under these particular experimental conditions was shown to be between 15 and 25 fish depending on individual variation within particular schools. Although these results are only valid within particular experimental constraints, they do suggest that fish schools in the wild could derive the benefit of enhanced ability to respond appropriately to environmentally relevant stimuli as a result of schooling behavior.

Host pose and intraspecific cleaning symbiosis in bluegill sunfish (*Lepomis macrochirus*)

Heather V. Roberts, *Walla Walla College, Dept. of Biology, College Place, Wash.*

Bluegill pose head down to be cleaned by smaller bluegill. This study examined the role of this perpendicular pose in the cleaning interaction by substituting models for the hosts, and noting the proximity of a juvenile bluegill using time-lapse photography (S-8). Response to the model was defined in terms of a theoretical cylinder (18 cm in diameter and 15 cm in length) projected horizontally from the model. The results indicate that the cleaner fish treat the models differently depending upon model angle (0, 45, or 90); specifically, they spend more time near models posing head down (90 from horizontal).

The effect of fish group size and cohesion on food detection

Lillian E. Roth & Benoni H. Seghers, *Dept. of Zoology, University of Western Ontario, London, Ontario, Canada.*

It has been postulated that schooling behavior is advantageous when fish are searching for food in a patchy environment. This hypothesis was tested using three species of fish differing in their tendency to school: the neon tetra (*Hyphessobrycon innesi*), zebra danio (*Brachydanio rerio*) and blind cave fish (*Anoptichthys jordani*). These were considered to be good, intermediate, and non-schoolers, respectively.

The experimental procedure was to allow single fish and groups of up to 25 fish to search for an inconspicuous 'patch' of food in a long narrow aquarium. The time for the single fish or the first and second fish in a group to detect the food was recorded. The results revealed that both schoolers and non-schoolers detected the food patch more rapidly than predicted by a simple model based on the time required for single fish to find the food. Nevertheless, in any particular group size schoolers were faster than non-schoolers. The neon tetras and zebra danios employed different foraging strategies which may be associated with differences in cohesion. There was also an indication that small group sizes may be optimal for foraging in a certain area size. Under natural conditions, however, foraging efficiency is only one of many potential selective forces acting on group size. In particular, the antipredator function of groups must be considered.

Effects of nest site concealment on hatching success, reproductive success, and paternal behavior of the threespine stickleback, *Gasterosteus aculeatus*

R. Craig Sargent, Joseph B. Gebler, Michael Brown, & George C. Williams, *Dept. of Ecology and Evolution, SUNY, Stony Brook, N.Y.*

In replicate wading pools, male sticklebacks actively competed for a limited number of clay flowerpots as nest sites. Males nesting in flowerpots spawned earlier and more often, had higher mean and lower variance for hatching success (number of fry per gram of eggs), and suffered fewer stolen fertilizations, nest raids, and territorial encounters than did males nesting outside flowerpots. Males nesting in flowerpots had a more temporally even distribution of fanning bouts and interfanning intervals; the duration of fanning bouts and interfanning intervals were more clumped for these males.

In replicate aquaria, one male nested on each side of a glass partition; both males spawned simultaneously. Males who had been given flowerpots had higher mean hatching success, fewer territorial encounters, and a more temporally even distribution of fanning bouts and interfanning intervals.

Thus it appears that nest site concealment, in the form of clay flowerpots, affected male hatching success, reproductive success, and parental behaviors, even when physical contact was prevented. These data suggest that a causal relationship may exist between the paternal fanning budget and the subsequent hatching success. Furthermore, these data suggest that females who spawn with males nesting in high concealment reduce the uncertainty of their hatching success, and possibly increase their absolute hatching success as well.

Spawning habits of the waccamaw darter

J.R. Shute, P.W. Shute, & D.G. Lindquist, *Dept. of Biology, University of North Carolina at Wilmington, Wilmington, N.C.*

The Waccamaw darter, *Etheostoma perlongum*, is one of three fish species endemic to Lake Waccamaw. The lake, which is located in southeastern North Carolina, is the largest of the shallow, sandy Carolina Bays.

Courtship and spawning behavior of *E. perlongum* has been observed on six different occasions in aquaria and once in Lake Waccamaw. Sexual dimorphism was well marked. The male was the larger individual and was much darker during the spawning act. His typical lateral 'W' markings were faded and appeared to be replaced by a series of 10 to 11 dark vertical bars. All median fins were noticeably darkened. A heavily pigmented area was evident in the anterior region of the spinous dorsal fin which was not present outside of the breeding season. The female merely exhibited more pronounced lateral 'W' markings, especially during the actual spawning sequence. Also, the genital papilla of the female was enlarged, making it more evident at this time.

The actual spawning sequence, which in most cases lasted between thirty minutes and one hour, consisted of male and female assuming an inverted position. This sequence occurred in a depression hollowed out by the male prior to spawning beneath a submerged stick. Between 50 and 150 adhesive eggs were deposited on the underside of the stick. After the spawning act was completed, the female was chased from the territory by the male, who remained in attendance of the eggs.

Because of the large numbers of eggs found under individual nests (with eggs in varying stages of development), it is likely that one male spawns with a number of females under the same nest throughout the spawning season.

Ethologists are environmental drop-outs

John Sprague, *Zoology Dept., University of Guelph, Guelph, Ontario, Canada.*

The United States is setting water quality criteria for potential pollutants. Data being used are mostly from lab tests on survival, growth, and reproduction of fish and invertebrates. Yet there is some tantalizing information that behavioral responses of fishes may be even more sensitive and thus of over-riding importance. Relatively few behaviorists have descended from their ivory towers to do such 'applied' work but they should; the field is being left largely to amateurs. Appropriate research could start with avoidance/preference reactions in the lab. Even more needed are studies of what reactions actually occur in nature when fish are faced with low-level pollution. Such research would be frustrating, messy, and probably ambiguous, yet it would be challenging and is urgently needed. There might even be funding available. I cannot understand why fish behaviorists have neglected pollution-related work.

Commensal spawning behavior of the rosefin shiner, *Notropis ardens*, in nests of longear sunfish, *Lepomis megalotis*

B.D. Steele, *West Virginia Dept. of Natural Resources, 1201 Greenbrier Street, Charleston, W.Va.*, & W.D. Pearson, *Water Resources Laboratory, University of Louisville, Louisville, Ky.*

Aspects of the reproduction of the rosefin shiner, *Notropis ardens*, and the longear sunfish, *Lepomis megalotis*, were studied in Barebone Creek, Trimble County, Kentucky. Special attention was paid to the unusual reproductive strategy of rosefin shiners. This was compared to the strategy reported for rosefin shiners in Steele's Run, Fayette County, Kentucky.

The rosefin shiner population in Barebone Creek was found to consist of four age classes with mature females first found in age class II. Rather than spawning in their own nests, as they are reported to do in Steele's Run, they were found to be commensal spawners in the nests of longear sunfish. The larval fishes in longear sunfish nests were as much as 85% rosefin shiner larvae. The spawning behavior of the rosefin shiners and longear sunfish was observed and in at least one instance a male longear sunfish and a male rosefin shiner were seen guarding the same nest.

Models were used in the field and in the laboratory to determine the reaction of each fish to the other species. The odors of longear sunfish milt and ova apparently attracted rosefin shiners to the longear sunfish nests.

Reproductive effort in unisexual and bisexual fishes: correlates with environmental stability

Roger Thibault, *Dept. of Biology, Bowling Green State University, Bowling Green, Ohio.*

In the genus *Poeciliopsis*, sperm-dependent unisexual fishes of hybrid origin, coexist with bisexual progenitor species. The mechanisms of unisexual adaptation and unisexual-bisexual coexistence in the rigorous environments of northwestern Mexico have been nurtured by several intriguing hypotheses. Investigations on differential fecundity and reproductive effort among unisexuals and bisexuals in three complexes are reported to contribute to the understanding of such mechanisms.

Over 4,775 fishes were dissected and ovarian contents tallied. Fecundity, a function of body size, was analyzed by multiple regression and analysis of covariance. Dependent variables were mean numbers of embryos per brood, mean numbers of mature ova and degenerate ova per female. Covariates included environment, month of collection and species. Additionally, over 1,600 females and their ovarian contents were independently dry-weighed to contrast unisexual-bisexual energy expenditure on reproduction.

For the *P. monacha* complex ($N = 2331$) bisexuals were more than twice as fecund as unisexuals at all four environments and at each of the six sampling months. Bisexuals correspondingly expend more effort on reproduction than unisexuals: about half of the female body weight in bisexuals to about one-fourth to one-third in unisexuals. Both unisexuals and bisexuals in the *P. monacha* complex exhibit increases in fecundity in response to increasing environmental stability.

For the *P. lucida* complex (N = 994), bisexuals again outproduced unisexual associates by two to one at each of three environments. Contrasting environments, however, did not expose differing fecundities with environmental stability. Unisexuals and bisexuals achieved maximum fecundities at opposite environmental spectra.

In the *P. latidens* complex, unisexuals outproduced bisexuals while expending less energy in reproduction. Although fecundity did not differ between the two environments, seasonality in reproduction differed among the two forms.

These results promote the concept of uniqueness of individual unisexual–bisexual complexes in similar environments. Hypotheses on mechanisms of adaptation and coexistence need be developed cautiously respecting the behavior and ecology of individual complexes.

Changes in the activity of a stimulus fish in the presence of a live conspecific

Don Townsend, *Dept. of Biology, University of New Brunswick, Fredericton, New Brunswick, Canada.*

An obvious concern with the use of live fish enclosed in clear tubes, bags, or boxes as visual stimuli when comparing various treatments is whether they provide relatively constant stimuli. For example, when such live fish stimuli are used to assess the effects of changes in interfish distance on agonistic behavior, is the stimulus fish behavior sufficiently variable to confound the interpretation of the results?

The swimming activity of confined live stimulus fish (*Tilapia mariae*) in the presence of a resident conspecific was determined over a range of interfish distances (2.4–20 cm) with the confined fish positioned at each of three above the substrate heights (2.4, 6.0, and 16.2 cm). This activity is described in terms of total time spent swimming, number of swimming bouts and average bout duration. Three different sizes of stimulus fish enclosures were used to permit varying amounts of movement.

Over the range of stimulus positions considered, for the two more confined stimulus situations a significant, but only low, proportion of the variation in confined fish activity could be attributed to stimulus position changes. In the third situation which provided the greatest freedom of movement, greater variation could be attributed to such effects.

Although a lack of response by the resident fish to the number of swimming bouts by the confined fish and the average bout duration was indicated under some conditions, the question of the nature and degree of response of the resident fish to the activity of the confined fish was still largely unanswered. However, particularly in situations where stimulus fish are provided with considerable freedom of movement, there exists at least the potential for stimulus activity variation that could affect the response to confined stimulus fish.

Color response in *Betta splendens*

Kathy Fusner Vail, *725 Dalrymple Rd., Apt. 5-C, Atlanta, Ga.*

Effects of color placement diagrams upon the display mechanisms of *Betta splendens* were determined. The presentation of scale diagrams with specific anatomical color combinations to individual *B. splendens* in a series of trials provided data. The diagrams without coloration, with only natural gill coloration and with gill and tail coloration elicited no response. Several combinations of color and placement caused a light aggressive display, while the diagrams with full natural coloration, with dorsal fins and spread coloration (tail) and with dorsal fins coloration alone provided the greatest degree of aggressive display recorded. In a related series of experiments, aggressive display responses were induced in *B. splendens* in order to determine the existence of behavioral patterns. Black and white photography was utilized to record these patterns and to relate individual display mechanisms to one another. The film sequence shows that display in *B. splendens* consists of simultaneous fin movement rather than movement of individual fins in sequence.

Contents

Proceedings of the 2nd biennial symposium on the Ethology and Behavioral Ecology of Fishes

Preface, by D.L.G. Noakes & J.A. Ward 5
A new stereophotographic technique for analyzing the three dimensional structure of fish schools, by L.M. Dill, R.L. Dunbrack & P.F. Major 7
Computer tools for the analysis of schooling, by M.J. Potel & R.J. Wassersug 15
Facultative schooling behavior in the spottail shiner (*Notropis hudsonius*): possible costs and benefits, by B.H. Seghers 21
Effects of alarm substance on schooling in the common shiner (*Notropis cornutus*, Cyprinidae), by E.J. Heczko & B.H. Seghers 25
Attraction of zebrafish, *Brachydanio rerio*, to isolated and partially purified chromatographic fractions, by F.D. Algranati & A. Perlmutter 31
Reduction of competition between bisexual and unisexual females of *Poecilia* in northeastern Mexico, by J.S. Balsano, K. Kucharski, E.J. Randle, E.M. Rasch & P.J. Monaco 39
Synchronized maturation and breeding in natural populations of *Xiphophorus variatus* (Poeciliidae), by R. Borowsky & J. Diffley 49
Maintenance of female mimicry as a reproductive strategy in bluegill sunfish (*Lepomis macrochirus*), by W.J. Dominey 59
Patterns of parental investment, dispersal and size among coral-reef fishes, by G.W. Barlow 65
The reproductive behaviour of *Aequidens vittatus* (Pisces, Cichlidae) in Surinam, South America, by M.H.A. Keenleyside & B.F. Bietz. 87
Reproductive tactics of the Asian cichlids of the genus *Etroplus* in Sri Lanka, by J.A. Ward & J.I. Samarakoon 95
Habitat segregation in a salt marsh among adult sticklebacks (Gasterosteidae), by J.P. Worgan & G.J. FitzGerald 105
Factors affecting fish diversity on a temperate reef, by J.S. Stephens, Jr. & K.E. Zerba 111

Abstracts 123

Second biennial symposium on the Ethology and Behavioral Ecology of Fishes, held at Illinois State University, Normal, Illinois 19-22 October 1979

a

b

c

d

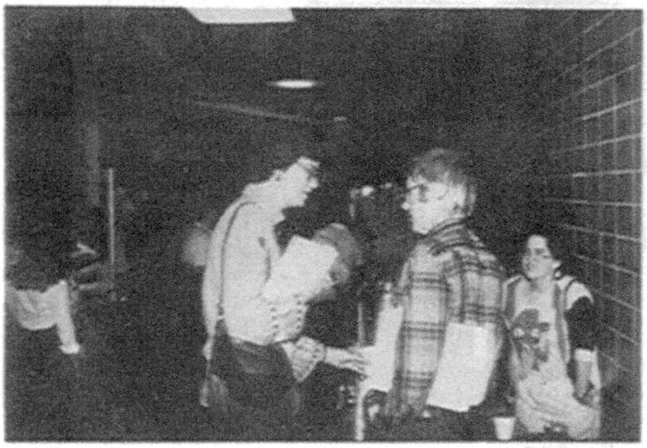

e

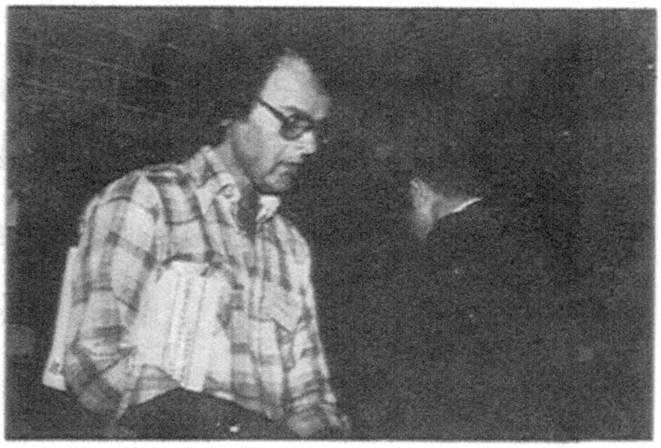

f

a. The organizational committee, from left to right: Ben Seghers, Bill Rowland, Luther Brown, Jeff Baylis, Mike Fitzsimmons, Chris Thoms. Mary Henry, Jack Ward and David Noakes (Richard Wyman was not present). *b.* Participants outside the auditorium for the plenary session, from left to right (in face view): Terry Lim, Margaret Matsui, George Barlow. *c.* Relaxing outdoors between papers, from left to right: Bill Rowland and Don Jensen, with Brian Partridge in the far background. *d.* Larry Dill, and Dr. and Mrs. Lars Hoglund in the living room of the Ewing Cultural Center at the wine and cheese reception opening the three-day conference. *e.* Diana Bauer, Daryle Waechter and others exchange points of view at one of the many coffee breaks. *f.* Jerry Downhower checks his notes before giving his paper in the 'Reproductive Tactics' symposium.

GPSR Compliance

The European Union's (EU) General Product Safety Regulation (GPSR) is a set of rules that requires consumer products to be safe and our obligations to ensure this.

If you have any concerns about our products, you can contact us on ProductSafety@springernature.com

In case Publisher is established outside the EU, the EU authorized representative is:

Springer Nature Customer Service Center GmbH
Europaplatz 3
69115 Heidelberg, Germany

Batch number: 10824412

Printed by Printforce, the Netherlands